Sustainable Development: Closing The Loop

Sustainable Development: Closing the Loop

Saket Kushwaha

Delve
PUBLISHING
www.delvepublishing.com

Sustainable Development: Closing the Loop

Saket Kushwaha

Delve Publishing
224 Shoreacres Road
Burlington, ON L7L 2H2
Canada
www.delvepublishing.com

Email: orders@arclereducation.com

ISBN: 978-1-77407-616-3 (Hardcover)

ABOUT THE AUTHOR

Prof. Saket Kushwaha, currently the Vice Chancellor of Rajiv Gandhi University is specialized in resource management and sustainable agriculture development. After his higher studies from Banaras Hindu University, he joined Abubakar Tafawa Balewa University (ATBU), Bauchi, Nigeria in 1993 and taught various courses on Agriculture Economics and Management at undergraduate, post graduate and Ph.D. level. Prof. Kushwaha rose to the rank of Professor at ATBU in the year 1999 and in 2006 he joined Bananas Hindu University (BHU), India as professor in agriculture economics and became the Vice Chancellor of Lalit Narayan Mithila University for one term 2014-2017. He has more than 100 publications in national and international journals of repute, supervised 24 Ph.D. students and authored 17 books / book chapters Prof Kushwaha is life member of 10 Professional Bodies and sits in the panel of editorial boards. Worked extensively in the field of Zero Emission Research Initiatives (ZERI) propagating the mission of sustainable development under the aegis of "Waste is Wealth" concept. He is also the recipient of 17 national and international

awards which includes award from Sulabh International Gold Medal in 2016 for sanitation management. 27 years of teaching, research, extension and community service experience with 23 years in administration. Handled over 10 national and International projects majorly funded by World Banks. Coordinated USAID project on Cowpea Research Support Programme (CRSP) in Nigeria for 10 years from 1996 to 2006. Working with NGOs and mega agriculture farms for Green Farm Planning.

TABLE OF CONTENTS

List of Figures .... *xi*

List of Tables .... *xiii*

List of Abbreviations .... *xv*

Preface .... *xvii*

Chapter 1 Fundamentals of Sustainable Development **1**

1.1. Introduction 2

1.2. Impacts of SD on Business and Industry 9

1.3. Implications For Farming and Agriculture 14

1.4. Advances In Sustainable Development (SD) 21

References 28

Chapter 2 Sustainable Development and Environment **39**

2.1. Introduction 40

2.2. Environmental Resources 40

2.3. Environmental Indicators 41

2.4. Environment and Sustainable Development (SD) 42

2.5. Environmental Interdependence 45

2.6. Problems Related to Environment And International Trade 46

2.7. Environment and Economic Growth, Trade-Off 47

2.8. Externalities Impact on Environment 48

2.9. Environmental Models 49

2.10. Immoderate Utilization of Resources 54

2.11. Developed Countries (DCS) Contribution in Environmental Pollution 55

2.12. Environmental Hazards Affecting Human Health 56

2.13. Environment and Market Incentives 58

References 60

Chapter 3 **Energy and Sustainable Development** **69**
3.1. Introduction 70
3.2. Energy and Social Issues 76
3.3. Energy, The Environment, and Health 80
3.4. Energy Security 85
3.5. Energy Resources 86
3.6. Energy End-Use Efficiency 88
3.7. Renewable Energy (RE) Technologies 91
3.8. Advanced Energy Technologies 92
References 100

Chapter 4 **Opportunities and Issues In Renewable Energy Technologies For Sustainable Development** **111**
4.1. Introduction 112
4.2. Socio-Cultural Barriers 112
4.3. Information and Awareness Hurdles 115
4.4. Market Fiascos and Economic Barriers 116
4.5. National and International and Strategies For Sustainable Development 119
4.6. Private, Local, and Nongovernmental Sustainable Development (SD) Initiatives 125
References 127

Chapter 5 **Economic Features of Sustainable Development** **143**
5.1. Introduction 144
5.2. Sustainable Growth 144
5.3. Environmental Economy 147
5.4. Socio-Economic Aspect of Sustainable Development (SD) 149
References 153

Chapter 6 **Social Aspects of Sustainable Development** **159**
6.1. Introduction 160
6.2. Sustainable Development (SD) Indicators 161
6.3. Social Sustainability 164
6.4. Social Aspect In Sustainable Development (SD) Of An Enterprise 166

References ... 171

Chapter 7 Future Prospects of Sustainable Development ... 175

7.1. Introduction ... 176

7.2. Human Well-Being And Competencies ... 177

7.3. Sustainable And Economies ... 178

7.4. Food Systems And The Patterns Of Nutrition ... 179

7.5. Energy De-Carbonization With Worldwide Access ... 180

7.6. Urban And Peri-Urban Development ... 181

7.7. Global Environmental Commons ... 182

7.8. Science For Ecological Development ... 183

7.9. Not Incremental Alteration And Transformation ... 183

References ... 185

Chapter 8 Issues In Sustainable Development and Their Potential Solutions ... 191

8.1. Introduction ... 192

8.2. The Necessity For Sustainable Development (SD) ... 193

8.3. The Origin of Sustainable Development (SD) ... 195

8.4. Hurdles In The Path of Sustainable Development (SD) ... 197

8.5. Potential Solutions ... 202

8.6. Creative Solutions For Sustainability ... 204

8.7. Areas of Innovative Solutions Requiring Focus ... 206

8.8. Role of The United Nations ... 209

References ... 214

Index ... 223

LIST OF FIGURES

Figure 1.1. Environmental management chart based on three fundamental spheres

Figure 1.2. The current state of the three spheres

Figure 1.3. Sustainable development

Figure 2.1. Environmental Kuznets curve

Figure 2.2. Estimates of global HIV infections (1980–2005)

Figure 2.3. ATC and MC curves for high-cost fuels

Figure 3.1. An example of the energy chain from extraction to services

Figure 3.2. Primary per capita energy consumption (commercial and non-commercial) by region, 1995

Figure 3.3. GDP and primary energy consumption in OECD countries, 1971–1996

Figure 3.4. Two energy intensity paths are shown for Japan and the United States, one based on total energy consumption from all sources and the other only on commercial energy

Figure 3.5. Average energy demand

Figure 3.6. Technology performance and costs improve with experience, and there is a pattern of such improvements common to many technologies

Figure 6.1. Sustainable development-social aspects model

Figure 6.2. Employees age structure

Figure 6.3. Employee's qualification structure

Figure 8.1. Keywords in twelve leading international media 1995–2001 (Nordic, 2002)

Figure 8.2. Trends of the world population

Figure 8.3. The term of sustainable development challenges

Figure 8.4. A tree being planted in 2012 by Grace, a chief member of the forest community in Ainabkoi, Kenya, and a farmer from the village of Kipilat

LIST OF TABLES

Table 3.1. World primary energy consumption, 1998

Table 3.2. Energy-related options to address social issues

Table 3.3. Environmental insults due to human activities by sector, the MID-1990s

Table 3.4. Current status and potential future costs of renewable energy technologies

Table 6.1. Total number of sustainable development's national indicators in European countries

Table 6.2. Sets of indicators utilized most frequently

Table 6.3. Social classifications and objectives in social indicator sets

Table 8.1. A brief outline of the remarks on the contorted theories

Table 8.2. A brief overview of the four competing points f view

Table 8.3. Aspects of societal conflict resolution

LIST OF ABBREVIATIONS

ADB	African Development Bank
ATC	average total cost curve
CO_2	carbon dioxide
DCs	developed countries
DNA	Designated National Authority
GDP	gross domestic product
GEO	global environmental outlook
GMOs	genetically modified organisms
GNP	gross national product
GPI	genuine progress indicator
ICLEI	International Council for Local Environmental Initiatives
IGCC	integrated gasifier combined cycle
IPBES	Intergovernmental Science-Policy Platform on Biodiversity and Ecosystem Services
IPCC	Intergovernmental Panel on Climate Change
IT	information technologies
IUCN	International Union for Conservation of Nature
LDCs	less developed countries
LWRs	light water reactors
MC	marginal cost
MDGs	millennium development goals
NSDS	National Sustainable Development Strategies
ODA	overseas development aid
PPP	purchasing power parities
RE	renewable energy
SD	sustainable development
SDGs	sustainable development goals

SSC	Stroller Chemical Co.
UDC	under-developing countries
UNCED	United Nations Conference on Environment and Development
UNEP	United Nations Environment Program
UNISDR	United Nations Office for Disaster Risk Reduction
WCED	World Commission on Environment and Development
WHO	World Health Organization

PREFACE

Sustainable development (SD) necessitates economic, social, and environmental policies to be planned and applied in a mutually strengthening way. This indicates a need for novel management cognition to improve coherence in policy and to enhance the role of information in the development and execution of policies along-with improvement in communication with business and civil society. SD must not be regarded as an extra obligation but a predominant principle, which administers the development procedures. SD is linked with the methods for the development of long-term activities of humans guaranteeing the harmony of economic, ecological, and social equilibrium. Sometimes, the risks associated with the violation of SD principles are of diverse kinds. Nevertheless, the consequences of various unsustainable activities of human beings hurt the ecosystem badly and may impact the survival of upcoming generations.

The underlying agenda of this book on SD is to provide detailed insight into the principles and prerequisites of SD. In order for us to achieve SD, it must be comprehensive, i.e., it should successfully balance social goals with environmental and economic perspectives. "Development" is actually more than just economic growth. The meanings of development can vary among different countries and even amongst individuals. However, it typically goes beyond the goal of amplified average income to encompass things like equity, health, freedom, education, and a safe environment.

There are eight chapters in this book "Sustainable Development-Closing the Loop." Basic knowledge of sustainability and development is necessary to grasp the underlying concepts of SD. Chapter 1 introduces the readers with fundamental concepts of sustainability and SD. Chapter 2 illustrates the relationship between the environment and the SD. Energy sustainability is a fundamental problem in the contemporary world. Chapter 3 contains a detailed discussion about the SD in terms of energy.

There are different kinds of renewable energy (RE) technologies that are replacing conventional energy technologies. Chapter 4 focuses on different sustainable energy solutions for achieving SD in the energy sector. Chapter 5 discusses the economics of SD. The overview of the different economic impacts of sustainability is given in the chapter. Social impacts of SD are also important in conjunction with economic and environmental impacts. Chapter 6 illustrates the underlying social impacts of SD. The future of earth heavily depends on the actions of human beings to combat global warming and climate change. Chapter 7 focuses on the future prospects of SD. Finally, Chapter 8 discusses the issues in SD and their potential solution.

Young students and learners of diverse age groups (including teachers, scientists, researchers, businessmen, and governments) will find this book exceptionally useful. This book can be used as a ready reference for all the topics related to sustainability and SD.

—Author

CHAPTER 1

Fundamentals of Sustainable Development

CONTENTS

1.1. Introduction 2
1.2. Impacts of SD on Business and Industry 9
1.3. Implications For Farming and Agriculture 14
1.4. Advances In Sustainable Development (SD) 21
References 28

1.1. INTRODUCTION

The idea of sustainable advancement (SD) emerges from another perspective, which sees the endurance, preceded with progress, and upkeep of the human network as reliant on the proceeded with wellbeing and feasibility of the earth's life support system. The expression "sustainability" gets from the Latin root *sustinere*, which intends to "under-hold" or hold up from underneath, suggesting strength and toughness after some time. Likewise, sustainability portrays a worldview that tries to secure the planet's life support systems to guarantee a life span for people and different species (Robert et al., 2005; Nilsson et al., 2016).

Economic improvement can be characterized as the procedure of key changes in our social systems and establishments required to accomplish sustainability. The expression "development" is condemned by some as connoting development. The free development of the "natural impression" of the human populace, characterized as the segment of the biosphere used to help human creation, utilization, and waste, is, obviously, eventually not sustainable. Others decipher "development" to mean advancement in social prosperity or improvement in personal satisfaction. Sustainable development (SD) is, accordingly, an antagonistic and intriguing idea (see "Sustainability in International Law"). To react to the challenge of discovering ways by which all individuals from the human family can live fulfilling lives inside the methods for nature (William Rees' definition), would require cooperative endeavors from an assortment of gifts: masterminds about society, researchers, and experts, business pioneers, ranchers, governments, and citizens (Hák et al., 2016; Le Blanc, 2015).

The major reason that supports the idea of SD is that the people groups of the world depend on their endurance on an environmental system that is both worldwide and limited. Subsequently, watching nature's limits is significant to forestall irreversible exhaustion of the life support systems. Up to this point, the idea of the earth as a limited framework was difficult to comprehend and pass on, for the earth had consistently appeared to be so immense and boundless. The advent of space travel brought awareness. Margaret Mead declared that the primary image of the earth as a little desolate blue ball in space, looking "vulnerable and requiring protection from the attacks of the technological man" gave force to the natural development. Canadian space traveler Roberta Bondar depicted our planet as "a fresh, splendid gem in space." Some space travelers noticed the rising smoke from the consuming of the Amazon timberland, and others the high-energy use bits of the world's surface truly sparkling around evening time. Every one of these images has given new visual importance to the illustration "spaceship earth," which was begat by Kenneth Boulding during the 1950s (Gupta and Vegelin, 2016; Stafford-Smith et al., 2017).

The image of the earth from space gave a first look at the "big picture of sustainability" since it indicated the limits and vulnerability of earth's life support systems and clarified the worldwide natural effects of human activities (Sachs, 2012; Gupta and Vegelin, 2016).

The worldwide environmental problem underscores the requirement for another way to deal with the knowledge that will broaden our capacity in critical thinking by looking at and watching the long term and total effects of different marvels (Mitchell, 1996, 2008; Griggs et al., 2013). The test is mind-boggling and multifaceted. John Robinson and Jon Tinker have built up a value system based meaning of sustainability as a creative, incorporated reaction to three "imperatives;" natural, monetary, and social:

- The "environmental imperative" is to stay inside the planetary biophysical conveying limit.
- The "economic imperative" is to guarantee and keep up satisfactory material ways of life for all individuals.
- The "social imperative" is to give social structures, including frameworks of administration, which successfully proliferate and support the qualities and culture individuals wish to live by.

Most sustainability scholars and professionals would grow the components of the social imperative to incorporate a major duty to both intergenerational and intragenerational value. A sustainable world must accommodate the fundamental needs surprisingly living today ("intragenerational value") without blocking the capacity of people in the future to meet their needs ("intergenerational value").

The technological capacity to see the earth from space has freed us from the myopic tendency to treat issues, including both natural and social ones, from a limited, restricted point of view. Contamination knows no political limits. Also, no nation can protect itself from social unrest, whatever its causes (Tueth, 2009; Gherasim and Tanase, 2012). This is one of the indelible lessons of September 11, 2001. With the present innovative advancement in interchanges, nobody can get away from the stun and ghastliness of seeing another kindred human in desolation anyplace on the planet. As Marshall McLuhan forecasted, microelectronics has made a sort of global village and has made visible the human condition in each side of the world, presently like never before. The new worldwide awareness has uncovered the multifaceted and urgency the idea of numerous cultural issues; and at any rate in idea, has connected each express, every area, each person, presently, and later on, to a similar sustainability challenge and fate. A mind-blowing entirety life support systems-including our social and financial systems-are comprehensively interconnected and related (Bossel, 1999; Glémain, 2011).

In 2001, the UN Secretary-General's report indicated that the total population would be around 8 billion by 2025 and 9.3 billion in 2050. This report additionally gave insights demonstrating that though worldwide food creation has kept on extending more quickly than the population in the previous decade, some horticultural practices have prompted natural disintegration. During this period, farming extension kept on infringing into woodlands, meadows, and wetlands. The pace of deforestation for the 1990s is 14.6 million ha for each year (Lélé, 1991; Costanza and Daly, 1992). This has been caused for the most part by the extension of agricultural practices and urbanization. It was assessed that a 17% increase in water supply would be required for nourishment developing in the following twenty years, squeezing the issue of water shortage. By 2025, 66% of the total populace could be living in countries with moderate or extreme water pressure. Right now, 815 million individuals on the planet are undernourished. Of these, an expected 777 million live in developing regions, 27 million in the transition countries, and 11 million in industrialized nations (Elkington, 1994; Cash et al., 2003).

Populace development combined with an expanded pace of resource utilization, and waste accumulation, will prompt natural decay, and land, nourishment, and water deficiencies. Resource courtesies unevenly caught by the rich and the poor will additionally enrapture countries and networks and make disastrous results. In the time of the 1990s, an expected 80% of total populace increment occurred in urban centers, yet the urban centers are not constantly sound spots. Urban centers have a sizeable underclass of individuals who are poor, wiped out, jobless, or exploited (see "Urbanization"). Jim Wolfensohn, the previous President of the World Bank called this dangerous situation of natural debasement and developing social disparity, a "time bomb" which, if we don't make a move currently, "could detonate in our kids' faces" (Bell et al., 2001; Bell and Myers, 2001).

The worldwide environment is experiencing noteworthy changes that are affecting society now and for the future, including environmental change, expanded waste and contamination, and the consumption of natural resources. Worldwide social issues of unemployment, poverty, violent conflict and disease have arrived at staggering levels and are in numerous regards getting more worse (Parris and Kates, 2003; Bansal, 2005).

Sustainable advancement is a reaction to the time bomb caution. The meaning of sustainable advancement gave by the Brundtland Commission in Our Common Future (1987) is "development that addresses the issues of the present without compromising the capacity of people in the future to address their issues." It is about how to prepare effectively international, national, community, and human intellectual resources by and large, towards taking care of issues that are, in general, corrupting the life support system, bargaining the suitability of nearby communities, and harming the wellbeing and prosperity of people. It is tied in with making a superior option in contrast to the system grinding away, because, given our present comprehension, the nothing new situation will carry extraordinary debacles to mankind, and is not manageable (Dincer, 2000; Sharpley, 2000).

Sustainable improvement premise revolutionizes the customary perspective on ecological administration, which sees the elements of society comprising of the transaction and interaction of spheres: financial, environmental, social, etc. Hence, likewise, natural administration is about making propels inside the ecological division (e.g., contamination observing and relief advancements) and in territories where the ecological intrigue is viewed as covering with other sectoral needs (for instance, work in resource extraction and processing industry, new lodging in rural areas). The chief disadvantages of this customary view are that, first, ecological concerns become a tradeoff in each arrangement. The quality and the general execution of the life support system turn into a minor idea in retrospect. Second, it neglects to recognize that numerous natural issues are fundamental and related. The piece of information to an answer may well lie in a basic change in perspectives, practices, and world view (Bell and Schwartzberg, 2001; Omer, 2008) (Figures 1.1 and 1.2).

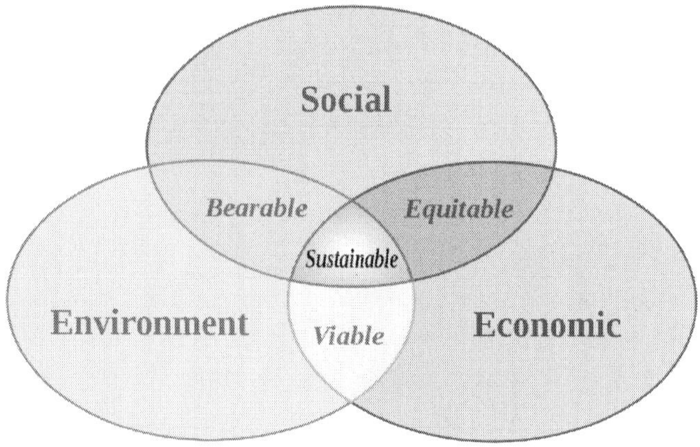

Figure 1.1: Environmental management chart based on three fundamental spheres.

Source: https://en.wikipedia.org/wiki/Natural_resource_economics.

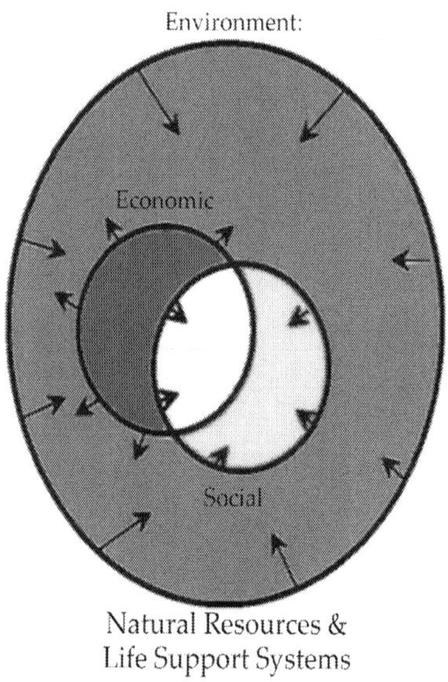

Figure 1.2: The current state of the three spheres.

Source: http://lsf-lst.ca/media/resources/en/cri/documents/David_Bell_and_Annie_Cheung_Introduction_to_Sustainable_Development_EOLSS.pdf.

In structuring an instrument for stimulating changes in the course of economical advancement, one must recognize the deficiencies of the current system and gain from experience. The organizations that make wealth in the new economy are changing from being asset-based to knowledge-based. Innovative change has empowered a shift in assembling creation from western industrialized countries to plants in Asia and Latin America. Work itself is being changed, driven by fast changes in innovation, international exchange, and financial rebuilding. Globalization increasingly uncovered all economies, yet particularly exporter nations, to the elements of world markets, with mixed outcomes. The financial atmosphere is changing as the role of business grows and that of government contracts. Indeed, even the meaning of riches itself must be adjusted (Cheung, 1997; Dolzer and Thesing, 2000). Mark Anielski utilizes the idea of "genuine wealth" which shows the etymological root of wealth which implies actually "the condition of well-being." Enhancements in genuine wealth cannot come to the detriment of the indigenous habitat. Nor can the abundance of the general public be estimated properly utilizing obsolete ideas like GNP. The way of sustainable improvement requires the utilization of measurements like the "genuine progress indicator" (GPI) (Figure 1.3).

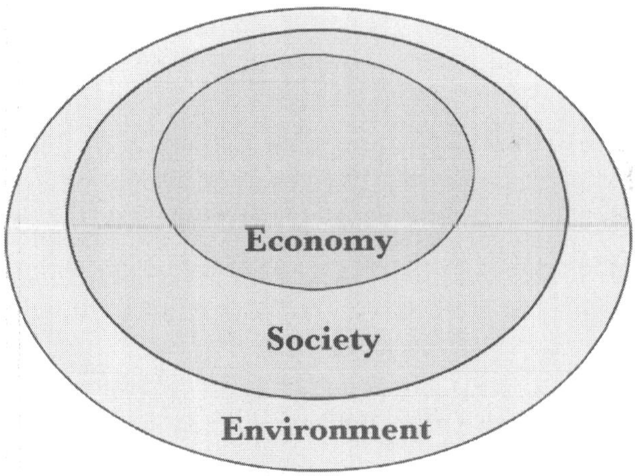

Figure 1.3: Sustainable development.

Source: https://sustainabilitypopulareducation.wordpress.com/2014/07/03/sustainability-if-its-everything-is-it-nothing/.

Advanced industrial societies' customary reaction to unpredictability has been to compartmentalize issues and manage issues independently, each inside its specific situation. Knowledge is sorted out in discrete trains inside scholastic institutions, cultural issues are separated among government departments, and abilities are distributed among professions and job classifications. Elites in these social orders have specific to develop expertise, however at the expense of losing a point of view on society all in all. Subsequently, separate societies have developed. A gap has created between individuals occupied with social policy and community; the science, innovation, and business networks; and those engaged with securing the earth. Every one of these isolations has its presumptions about society, in light of various encounters, particular dialects and different translations about the circumstances and logical results of normal concerns (Foster, 2001; Omer, 2008).

These polarities become obvious in conversations about government needs, especially in the wealthier nations. Finance organizations talk about diminishing duties, obligations, shortfalls, and unnecessary open spending. Social offices center around value issues; dangers to the burdened, unfulfilled needs, and harm to the social infrastructure of wellbeing, government assistance, and training because of cuts in broad daylight spending. Ecological agencies focus on issues of environmental change, water, and air quality, and biological system conservation. To all goals and purposes, these dissimilar gatherings inside government and society seem to live in various universes, and frequently see their destinies as free of each other. They neglect to perceive how the choices taken in one circle or division will at last effect on all the others (Orr, 1991; Prakash, 2000).

In schematic terms, the monetary, social, and environmental spheres of the here and now look like three spheres of changing sizes. The ecological sphere is as yet the biggest, in which both the social and monetary spheres arrange and connect. The monetary sphere is developing in size; the social sphere and the ecological sphere are both contracting. It might be legitimate to foresee that the arrangement of the spheres is moving towards a pattern of concentric circles: ecological, social, and financial in the climbing request of size. Be that as it may, this cannot be a reasonable expectation, because, at one point in the change when the constraints of nature will become to, the environment of mankind will never again be practical, the social circle will fall to pieces and the economy will come to a standstill. This is the disaster that one must stand up to and endeavor to stay away from (Stave and Durandetta, 2000; Unctad, 2000a, b).

Researchers, specialists, and government officials have accused our present careless disposition towards our life support system on a large group of weaknesses:

- The absence of comprehensive reasoning and practices-compartmentalization of single control considering and institutional courses of action.
- The absence of global regulatory establishments and lawful arrangements.
- The absence of progressing intergovernmental exchange, dualism in our power structure and the disregard of women's activist concerns spoke to in realpolitik.
- The brief time-frame outlines and powerless political will of elected governments.
- The absence of coordination among the divisions of government.
- The absence of tenacity and capacity in our regulatory organizations.
- The absence of understanding among the populace.
- The shortcoming of the public will during bad economic occasions.
- The transient skylines of business.
- The lapse in public-spirited activities.

Disconnection and "missing connections" inside and between the domains of standards and practices are viewed as the principal shortcomings in the present framework (WBCSD and UNDEP, 1999; United Nations Press Release, 2000a, b).

1.2. IMPACTS OF SD ON BUSINESS AND INDUSTRY

Ventures manufacture consumer merchandise that satisfies the needs of regular day to day existence. Organizations offer types of assistance that encourage a two-way process, going about as to some degree a questioner between the customer and the maker. Business and industry have a fundamental impact on accomplishing the cutting edge way of life. They likewise make occupations for some individuals and structure the significant segment of our economy. The improvement of ventures and organizations is firmly identified with the elements of the customer advertise, which is dictated by the developing number of purchasers, the buyers' needs, and their desire to purchase new items (Lawn, 2001; Cobbinah et al., 2015).

The production of merchandise and ventures by business and industry likewise relies upon a consistent stock of raw materials-including energy, water, nourishment, mineral resources, etc-or, at the end of the day, the accessibility of natural capital. Characteristic capital is more than the industry's stock. It typifies an automatic procedure of energizing and recovery, which is empowered by the interaction and presence of a system of natural components important for this procedure (Timur and Getz, 2009). A similar idea of self-guideline applies likewise to the human natural habitat. On the off chance that specific components of nature are by and large exorbitantly extricated, the adjustment in a critical position of the existing life support systems may thus additionally steamed the human living space, influencing the wellbeing and prosperity of mankind. In this manner, the over-reaping of nature for assembling materials will eventually affect the personal satisfaction of the human community. For instance, the enormous scope, exploitative gathering of timber from normal backwoods likewise annihilates homes of creatures, slaughters the undergrowth and other vegetation, expels the repository of photosynthesis for clean air, and upsets the energizing regions of the underground water system (Jolink and Niesten, 2015).

Meanwhile, manufacturing processes release wastes. New product offerings eliminate old ones and transform obsolete things into the trash. Both industrial waste and trash are strange to the earth and its natural cycles. A considerable lot of the man-made materials, for example, spent atomic fuel or plastics, are not effectively biodegradable. Atomic waste, for example, plutonium 239 has a half-existence of more than 20,000 years. That is to say, it takes more than 20,000 years for just a single portion of the radioactivity present in the matter to disintegrate. Along these lines, the removal of atomic waste represents the sort of mind-boggling and critical issues that must be handled by advanced technology and innovation, and by correspondence, especially with the people in the future who are to acquire the issues from us (United Nations Population Fund, 2001; Schaper, 2002).

If one sees the production and consumption process from a social point of view, the more services and goods individuals consume, the more noteworthy material solace is accomplished, and the more individuals are being utilized to make the products and offer the types of assistance. In like manner, the proceeded with the activity of the production line is significant for supporting work; it conveys a critical social government assistance work. Moreover, when new items are created in quick progression, there are more item choices, and more amassing of inventory in the stock. To keep up equalization in stock size, and to dispose of the old stock, vendors bring down the costs of wares. Serious valuing and the fast turnover of items, for the most part, make the materialistic way of life progressively reasonable. All things considered, the working conditions of workers, inconsistent wages for female specialists and illicit settlers, long working hours, poor indoor workplaces, and industrial safety risks are likewise factors that undermine the nature of social life (Eden, 1994; United Nations Economic and Social Council, 2001).

For the economy, the higher the estimation of services and goods, the more significant the level of profits that can be made by the business sector and industry. This could prompt more noteworthy reserve funds and interest in the economy. Commonly, there are three different ways to raise benefits: an expansion in cost, an increment in yield, and a decrease in assembling costs. The expansion in yield relies upon the extension of the market and hidden that, the development of consumer demand. This is frequently modified by the adjustment in demands (for instance, pocket-size adding machines instead of the utilization of logarithm tables as an arithmetic tool) and design trends (the diverse dress lengths of the style season). Business promoting, a flourishing global business, assumes a noteworthy role in manipulating and stimulating the inclinations of current utilization (WCED, 1987; US Office of the Press Secretary, 2000a, b).

The most widely recognized practices for product creation costs have been down to lessen work and raw material expenses. The previous alludes to the act of utilizing modest non-unionized workers, subcontracting work to cheap labor in the developing nations, or by supplanting workers with computerization and the clever utilization of innovations. The last is ordinarily accomplished by the concealment of natural resource costs, and by barring from the monetary records the estimation of the life support system (Robert et al., 2005; Wackernagel and Rees, 1996).

The more we devour, the more merchandise is made, the more natural resources are changed over into crude materials, the more things are rendered repetitive and are transformed into the trash that litters the earth, the more waste is released into the life support systems. This is the natural effect of production and consumption activities. The measure of waste produced in countries like Canada and the United States is faltering. The normal American devours twenty tons of raw materials every year. The all-out yearly waste produced in the United States would fill a caravan of dump trucks sufficiently long to fold over the earth multiple times and reach most of the way to the moon (Holthus and Council, 1999; Welford, 2013).

As of late, the dramatic development in worldwide exchange and venture, encouraged by the progression in media transmission, transportation, and computerization, has encouraged the creation of a worldwide scale division of work (World Trade Organization, 2001). The incredible increment in worldwide exchange and interest as of late has been a significant promoter of the geographic centralization of monetary activities. As per the World Trade Organization, in the year 2000, the absolute world incentive for export products and ventures was US$7.6 trillion. Regardless of the much, more slow economic development contrasted and the earlier year, in 2001 there was as yet a normal of 2% development in the worldwide economy. The estimation of world product exports came to US$6.19 trillion out of 2000, which is more than a hundred times the estimation of US$58 billion every 1948. Meanwhile, World Investment Report 2001 kept up additionally that the world foreign direct venture inflows grew twenty-two overlays from US$57 billion out of 1982 to US$1.271 trillion out of 2000.

Exchange the past, related to the military-political power of colonization, has extended the inventory of materials and the conveying limit of numerous countries: Great Britain France, the Netherlands, Japan, etc. Notwithstanding, the hardship on the indigenous individuals and the loss of characteristic courtesies were scarcely represented in these experiences. Presently, we start to have the apparatuses and the ethical basis to survey our exhibition on a worldwide scale (Azapagic, 2004; Shiyi, 2009).

Also, as per William Rees' figuring, the total eco-impression of the present total populace of 6 billion is as of now 20 to 30% bigger than the size of the earth. It implies that the human "load" has effectively overshot the long term conveying limit of the planet; or the pace of human utilization is more noteworthy than what the earth is fit for delivering uncertainly, without making some misfortune the world's own recovery ability (Smith, 1995; Holt, 2005).

An ongoing World Trade Organization (WTO) report anticipated the total populace to cross the 7 billion imprints in 2013. It has taken just twelve years to include the last billion people. This populace pattern, joined with the total populace's unreasonable utilization, is anticipating a miserable image of resource weariness that may happen not very far later on. Garrett Hardin's situation of resource over-abuse by the congestion of individuals, who do not fret about a need to deal with normal or shared resources, strikes a reverberating harmony.

The discussion that Hardin started about how to manage asset exhaustion issues that emerge from the misuse of "common-pool resources" by a relentless public, is pertinent and edifying to the challenge of resource the executives for sustainability. Hardin's recommendations incorporate government administrative measures: authoritative law that constrains capable conduct (e.g., environmental protection laws and guidelines), and disciplinary market mechanisms (e.g., "green" charges).

The conversations that have been created by Hardin's Tragedy of the Commons relate to the subject of the requirement for, and the roles that are played by, the different levels and kinds of expert in managing and pressuring consistency in individual and group activities (Solow, 1986; Common and Perrings, 1992).

The sustainable improvement point of view encouraged by the Earth Summit is one that recognizes all parts of the reality that has been featured, and that sustainable advancement cannot be accomplished without central changes in the manner communities deliver and consume. Three significant territories for business and industry activities have been recognized:

- Improvement in production practices;
- Designing for long-term utilization of the item; and
- Reducing the negative effects of transportation of products and material.

As far, as progress underway practices, the SD challenge includes an increasingly capable, successful, and productive application/use of resources, concerning both natural and human resources. There are incredible points of interest to be picked up in cultivating a feeling of correspondence and common enthusiasm for the workplace (Gibbs, 1994; Martin et al., 1998). Beside better work connections, better workplace conditions are essential for the development of a sense of responsibility; the sort of social standard that will have a positive effect on the authorization of the guideline and the shrewd utilization of assets in the creation procedure, including innovations, raw materials, energy, and local labor. Labor unions and laborers are significant accomplices in the change of the business and business areas towards a sustainability ethic (Ekpo, 2012; Prud'homme and Raymond, 2013).

To decouple economic development from ecological debasement through "dematerialization" (a decrease in the throughput of energy and material resources per unit of production) has been the principal guarantee of the business sector and industry in the light of the challenge of sustainable advancement. Reducing (ideally eliminating) waste by embracing a "support to support" way to deal with production expects to mimic the "closed-loop" cycles found in nature (Elkington, 1994; Sadorsky, 2014). Cleaner reduction and production of contamination, and the more productive utilization of natural resources ("eco-effectiveness") are the focal accentuations (Montana and Nenide, 2008). As of late, there has been the presentation of various financial instruments, including a growing range of natural taxes and charges that are coordinated into economic approaches of developed and developing nations; the tradable grant framework to control contamination discharges, water use, fisheries, and land improvement rights; store discount conspires in solid waste management; and rebelliousness expenses and execution bonds. These are the disciplinary market components that are ventured to be powerful related to authoritative laws forced by the government (Meister and Japp, 1998).

1.3. IMPLICATIONS FOR FARMING AND AGRICULTURE

As indicated by The State of Food Insecurity in the World 2001, an expected 815 million individuals on the planet are undernourished. Of this aggregate, an expected 777 million live in developing regions 27 million in transition countries, and 11 million in industrialized nations. Meanwhile, the Worldwatch Institute announced that there is the same number of overweight as deprived individuals on the planet. 80% of the world's hungry youngsters are living in nations that appreciate a nourishment excess. So while malnourishment is unequivocally connected to the state of poverty, in numerous cases the issue is not a flat out insufficiency in nourishment yet social imbalance. As Gandhi once stated, "There is enough for everybody's need, but not for everyone's greed."

The significant goal for sustainable cultivating and agribusiness is to build nourishment creation in a production manner and to improve nourishment security. During the 1990s, with the development of agrarian land, improvement in the water system concentrated cultivating advancements, utilization of new seed types, and alternative cultivating practices, the pace of worldwide nourishment creation became quicker than the populace. Tragically, this pattern has been hindered as of late by unfavorable climate conditions, for example, dry spell in West and Central Asia, and East Africa, floods in East Asia, a progression of catastrophic events in Central America, and wars (Berno and Bricker, 2001; Righi et al., 2011).

At the World Food Summit in 1996, world pioneers communicated a promise to accomplishing nourishment security for all and eliminating hunger in all nations, with a halfway objective of decreasing the number of undernourished individuals on the planet to a large portion of the present level by 2015. Since 66% of the world's poor populace live in rural zones, and participate in horticultural activities for nourishment and income, helping these individuals to accomplish upgrades in farming profitability will be a compelling method to help annihilate yearning and destitution. Supportable horticulture and cultivating, thus, must think about how to carry nourishment and advancement to the poor provincial networks of the world, just as how to improve the overall nourishment production in an earth sound way (Martinelli et al., 2010; Juana et al., 2013).

Numerous issues in the cultivating and horticulture division are connected to different worries with social, financial, biological, and innovative implications including, food security and poverty; gender equity; environmental change; climate change and venture; biotechnology; market access and trade liberalization effects; organic agriculture; biotechnology; knowledge transfer and information access. Also, farming practices and conditions contrast drastically between developing nations and developed nations. At the Commission on SD discussion on supportable farming, a delegate from South Africa remarked, "developed nations talk about horticulture as a way of life, though for developing nations it is a type of vocation."

Agricultural efficiency is identified with the size and appropriateness of land for crop development. The global landmass is restricted, and land that appreciates appropriate conditions for horticulture is topographically explicit. As a matter of convenience frequently markets and human settlement is set up inside driving separation to these prime farmlands, yet progressively urban spread is swarming out the farmlands (Scopel et al., 2013).

With more individuals moving to urban zones, most of the total populace got urban in 1999. Urban infringement into the neighboring rural land frequently prompts loss of ranch land to other conflicting and competing for land use and demolition of soil profitability for the rest of the farmland. One of the ways for the cultivating community to keep up profitability is to use increasingly natural woods, changing over nature to farms. The pace of deforestation all around during the 1990s was assessed at 14.6 million ha for every year, for the most part in tropical nations. The loss in woodlands, fields, wetlands, and so forth additionally diminishes the characteristic natural surroundings, biodiversity, just as the working of other ecological merchandise and services (Pretty, 1995).

Different ways of improving homestead profitability have met with mixed reactions. The development of the water system assists with broadening growing seasons and permits more land, which is, in any case, thought about unreasonably dry for cultivating, to be utilized for horticultural production. Water system plans become increasingly more sought after as degradation of dry land keeps on expanding (Schindler et al., 2015).

Internationally, desertification, an aftereffect of human activities and atmosphere changeability, has influenced around one-fourth of the all-out land region of the world, or around 70% of all drylands. With the expansion of built regions, changes in water catchment for underground water, and the artificial channeling of water system schemes, drylands do not recoup as quick as they used to from atmosphere inconstancy, for example, a long dry season. In this way dry land becomes desert.

Worldwide water reserve stays about the equivalent after some time, yet the extension in flooded horticulture and municipal and industrial utilization of water has made water pressure. By and by around 70% of the world's freshwater resources are utilized by farming at the same time, even with improved water system schemes; just 30% of this sum is taken up by harvests and plants. Evaluations demonstrate that in the following two decades, around 17% more water will be required to grow food for individuals living in developing nations, and all-out water use will increment by 40% (Rigby et al., 2001; Muzari et al., 2012).

Serious cultivating advancements are not without negative effects. Water system wastes water to dissipation, and fuels the issue of salinization, the aggregation of salt in the upper layer of soil because of poor waste. The act of applying synthetic fertilizers, instead of progressively conventional strategies including the utilization of compost and manure, is keeping the dirt from humus (which is genuinely necessary for keeping up the dirt structure), and dirtying underground water. Soil degradation has influenced in any event 2 billion hectares or as much as 66% of the world's agrarian terrains. Pesticide applied to crops has been discovered aggregated in nature, and rose in breast milk. Pesticide exposure is connected to a developing rundown of wellbeing issues: birth defects, cancer, reproductive harm, developmental, and neurological toxicity, disruption of the endocrine system, etc. The utilization of substantial hardware instead of farm laborers has prompted changes in farm size, relative abatement in the enlisting of homestead workers, and for the most part, a considerably more money escalated activity that depends on different types of sponsorships. Likewise, 70 to 80% of deforestation overall was because of huge, mechanized agriculture (Mutekwa, 2009).

By and by there are vivified banters over the utilization of genetically modified organisms (GMOs) in horticulture. As per the Worldwatch Institute, since 1996, the development of hereditarily engineered harvests has grown twenty-three overlays, and now the world has 39.9 million hectares of farmland dedicated to this kind of yield. These new seeds that are planted have been hereditarily modified so the harvests are impervious to the herbicide, or are intended to have the option to utilize the herbicide spray delivered by the dirt bacterium (Bt-crops) or to do both. These harvests are intended to streamline pest control by enhancing the impacts of herbicide spraying and decreasing the requirement for weeding (Van de Flirt and Braun, 2002; Paul and Vogl, 2011).

As indicated by the Worldwatch Institute report, in 1999, 71% of the worldwide transgenic region (territory that is assigned for developing hereditarily modified harvests) was planted with herbicide-resistant varieties of cotton, soy, corn, and canola, and 22% was planted with Bt-cotton and bt-corn. The hereditarily modified corn and soy were for taking care of animals, canola was for oil and processed food use, and the cotton delivered was for fiber and oil.

Reactions of rural biotechnology focus on two fundamental concerns:

- In the first place, the heading of biotechnology research can compound inequality; the new GMO seeds are not especially valuable for small-scale farms, or subsistence cultivating, where weeding is finished by hand. For enormous cultivating tasks, the disposal of weed is done usually by procured hands. When weeding gets pointless, numerous workers will be out of work (Despite what might be expected, one would not have any desire to get workers to work fields where chemical toxins are active). The GMO crops referenced were intended to be completely reliant on biocide, focusing on the cultivating business, and not especially given the poor third world farmers. Except if GMO seeds are intended for helping poor farmers in overcoming their horticultural issues, for example, the issues of high salt substance, dryness, and low supplement content in the dirt, biotechnology research will not help third-world farmers. Also, the most recent biotechnology improvement is producing an arrangement of protected germplasm. The spreading of GMO seeds from the transgenic territory to the contiguous wild and homesteads has been of worry to certain researchers. The individuals who do not wish to buy into the new seeds may incidentally discover them flourishing in their farms. In this manner, the matter of patented seeds and their access can be a wellspring of legitimate wrangling. Also, it may even affect the traditional method for cultivating and risk poor farmers' occupation (Ma et al., 2009; Orsini et al., 2013).

- Second, while the long term health hazard for human utilization of hereditarily modified yields stays obscure, there are different sides to the discussion on whether hereditarily modified food brings more prominent health dangers. A few researchers have contended that hereditary change of yields has been accomplished for a considerable length of time by farmers, who utilized traditional techniques for selection and hybridization. The new hereditary tools of biotechnology, as per the researchers, may make results that are increasingly unsurprising and more secure. Then again, adversaries of biotechnology have been demanding progressively longitudinal tests, away from test outcomes, and for the most part increasingly public input to choices concerning the business development of GMO crops (Van der Werf and Petit, 2002; Davis et al., 2012).

There is additionally a natural cultivating development in the agrarian network. Right now in the US, about 1.1 million ha of farmlands utilize natural cultivating management strategies. Natural cultivating suggests a dismissal of the utilization of chemicals, for example, pesticides, herbicides, and chemical fertilizers, which may wind up in the contaminate and produce the biological system. Rather, natural cultivating utilizes manure and compost, and "green manure" crops, (e.g., clovers, and ryegrass) as natural fertilizers, and worms and useful insects to make an appropriate developing condition for the yields. There are by and by more than 10,000 confirmed natural farmers in the US. Moreover, standardized labeling of natural nourishment to give better customer data is presently not too far off (Röling and Wagemakers, 1998; Jordan et al., 2007).

Economical agrarian practices must be created with energy about the rising pattern in worldwide utilization, and a potential descending pattern for arable land accessibility and soil profitability. As of late, the cultivating community is additionally worried by the changing climate example of long dry spells, flooding, and excessively warm or too chilly temperatures in numerous pieces of the world, that are influencing the growing conditions for some harvests. For sustainable food production, agrarian advancement will rely upon the improvement of cultivating strategies to monitor soil and water, to expend less fuel and less poisonous substances. Protection and eco-effectiveness should likewise be polished alongside better land the board to decrease the loss of land parcels to urban settlement, and to harm by flooding or landslides (Rigby et al., 2001; Lichtfouse et al., 2009).

Cultivating on the planet is directed in an extremely scattered manner. There are numerous decision-makers and actors (men, ladies, youngsters who take part in subsistence cultivating, commercial farmers, financial specialists, researchers, scholastics, legislators), who work freely however together they decide the food security of our future. Some entirely proficient ranchers use state of the art scientific innovations, and some utilize traditional information and increase awesome outcomes. Nonetheless, 66% of the world's agricultural lands are influenced by soil degradation (De Bon et al., 2010).

There is a need to keep moving for building a progressively foolproof framework to convey food security. A decent data arrangement that encourages the sharing of cultivating information is important to help welcome everybody forward-thinking on the most proficient method to amplify farm efficiency, without making further damage to the delicate farmlands and the environment. The adoption of the precautionary standard is significant for settling on astute choices that preempt harm before it happens. Be that as it may, the awareness of practicing alert must be gotten from a good information base and experience. There has been some conversation about the documentation of indigenous information to sell as a product for supporting rural improvement in the developing nations. Private sector biotechnology investigate is additionally producing GMOs as business products. A regulatory system will maybe be important to help encourage the trading of information and intelligence for sustainable horticultural advancement, in case of fast privatization of information (D'sSouza and Ikerd, 1996; Zahm et al., 2008).

Sustainable horticulture and cultivating must start with a superior comprehension of the general worldwide land resources, and will rely upon key arranging that is sympathetic to a bioregional point of view. It is likewise about attempting to advance the efficiency of farmland utilizing regenerative and sustainable horticultural practices. The world's ecological assets are unevenly circulated and area explicit. In specific conditions, attempting to accomplish food security for a region where the preconditions for successful cultivating do not exist, is negligible, and probably will not be attractive from a biological perspective. So also, if the guarantee of biotechnology is so remote from conveying the products that give poor people and hungry a superior occupation, and give everybody of us a superior dietary status and great wellbeing, the goals of biotechnology must be returned to. Besides, the principles for biotechnology research and application, either excessively stringent or excessively free, can hurt the possibility of nourishment security later on. Biotechnology offers new open doors however; one must proceed with caution (Smith and McDonald, 1998).

Farming harvests are a commodity of the agribusinesses. Commercial crops are additionally part of the natural framework. Hundreds of years of development of land and, explicitly, the most recent forty years of serious cultivating, have enormously affected the range, conveyance, and thickness of wild plants and creatures. Some arable weeds, which adjusted well to rural practice before, are presently viewed as imperiled, because of the utilization of pesticides and composts on farms. Any new agrarian advancement, or hereditary change of organisms, must focus on this and add to taking care of the issue. By a similar token, agri-business should in this manner be wary of any ecological effect that its activities may produce (Smith and McDonald, 1998; Zander and Kächele, 1999).

Farming is not just a commodity; it is likewise a job for some, who live in the rural communities of the developing nations, where hunger and poverty are uncontrolled. In any case, it must be noticed that improving horticultural profitability is not the main switch to get the impeded provincial populace out of hunger and poverty. Additionally required are an expanded measure of food being produced and traded, the ascent in income level of poor people, and a more noteworthy extent of investment going to agribusiness and rural improvement. There are 150 million underweight youngsters in the developing world (Altieri et al., 2012).

Ladies who consume a lot of energy on cultivating got the least relative measure of calories in the diet. The current shortcomings in society and rural institutions that block advancement and progress must be tended to. Specifically, ladies' job in rustic advancement and horticultural profitability should be underscored. An incredible number of ladies in developing nations are farmers just as meal providers for children. Recognizing ladies' equivalent status, permitting them to have equivalent access to training and data, to land rights, and money related help is significant for helping them to turn out to be better farmers and better guardians of nourishment. Any arranging methodology that neglects to address the criticalness of ladies as accomplices being developed and agrarian change will be neither powerful nor moral.

Accomplishing food security in any region requires the contriving of a vital way to deal with advancement by the nearby community. This advancement plan must consider how to best catch the domestic food security potential, with a view to not trading off at all the worldwide stock of farmlands and long term prospects for food production. The choice concerning whether to produce food locally or to rely upon food imports would likewise be determined on the near valuing of nourishment crops against different wares. The structure of international trading systems should reflect the objective of worldwide food security, so investment systems and lawmakers should likewise hold hands with a look into researchers, consumers, and farmers to seek after this goal (Dogliotti et al., 2005).

1.4. ADVANCES IN SUSTAINABLE DEVELOPMENT (SD)

Viewing the earth from space gave us an unmistakable image of the entire global framework. It has increased our mindfulness about the world's delicacy, and the systemic linkages that structure our life support system. Innovation has given us the apparatus to watch, screen, and evaluate the physical, biological, and chemical parts of progress in the natural habitat, and has helped us to find out about the different degrees of the effect that human exercises have on nature. Satellite technologies, among other things, have empowered us to watch the world's changing procedures all things considered and, subsequently, have broadened our exploration limit both in scale and in scope (Scholz et al., 2012; Hörisch, 2016). Time-series information created from remote-detecting and in situ observations have revealed to us a lot about almost every part of earth's biological systems, including the seas, the ozone layer, the upper air, worldwide woods inclusion, sea science, etc. In 2001, the Intergovernmental Panel on Climate Change (IPCC) gave the following report:

1. An expanding assemblage of observations gives an aggregate image of a warming world and different changes in the atmosphere framework.
   i. The worldwide normal surface temperature has expanded over the twentieth century by about 0.6°C.
   ii. Temperatures have ascended during the previous four decades in the most minimal 8 km of the air.
   iii. Snow spread and ice extent has diminished.
   iv. Global average ocean level has risen and sea heat content has expanded.
   v. Changes have additionally happened in other significant parts of the atmosphere, for instance: redistribution of precipitation, progressively intense and extraordinary dry spells in parts of Asia and Africa.
   vi. Some significant parts of the atmosphere show up not to have changed, for instance: a couple of regions of the globe, mostly over certain pieces of the southern Hemisphere seas, and parts of Antarctica, have not gotten hotter; no fundamental changes in the recurrence of tornadoes, thunder days or hail occasions.
2. Emission of aerosols and greenhouse gases because of human exercises keep on modifying the environment in manners that are relied upon to influence the atmosphere framework.
3. Confidence in the capacity of models to extend the future atmosphere has expanded.
4. There is new and stronger proof that the majority of the warming observed throughout the most recent fifty years is owing to human activities.
5. Human impacts will keep on changing environmental arrangements all through the twenty-first century.
6. Global normal temperature and ocean level are anticipated to ascend under all situations utilized (the IPCC Special Report on Emission Scenarios).
7. Anthropogenic environmental change will endure for a long time.

Our new information on the worldwide framework is a result of advances in geomatics, worldwide observation technologies, and coordinated research plan. The "Integrated Global Observing Strategy," uniting significant satellite and surface-based frameworks for worldwide natural perceptions of the air, seas, and land, has made conceivable the age of valuable worldwide information all the time (Ferrer-Balas et al., 2004; Skerlos et al., 2012). This has helped with planning woodland the management procedures, in propelling initiatives for the assurance of coral reefs, in bringing issues to light about the conceivable social, economic, and political effects that could be brought about by, for instance, the ascent of worldwide surface temperature; changes in precipitation magnitude, geographical, and intensity dissemination; and the vanishing of coastal areas. Innovations have improved transnational talk, for example, that enveloped by the United Nations Framework Convention on Climate Change (International Council for Local Environmental Initiatives (ICLEI), CAG Consultants and the Division for Sustainable Development of UNDESA, 1999).

New systems of appraisal and assessment of progress toward sustainability have likewise been received by the sociologies and applied to gauge cultural improvement. As of late, significant advancement has been made in the improvement of execution pointers in the social, normal resource, economic, and institutional spaces. These indicators articulate the states of a dynamic society contain fundamental foundation data for arranging and add to the announcement of the executive's objectives and benchmarks and the definition of legitimate contentions (Brumano et al., 2016). They encourage efficient examination of progress after some time and across communities. Possibly, they can be fundamental estimations of policy performance, giving proof of progress or failure. Also, they are specialized tools for use in expansive based consultation, and for decision making (Lerner, 1992; Wu et al., 2013).

As of late, the Commission on SD has been dynamic in creating "Indicators of SD," with the point of utilizing them as proportions of progress in networks far and wide. A significant deterrent, in any case, has been that few out of every odd nation have the resources fundamental for far-reaching information assortment (Rosen, 2019). Different concerns are the reliability of the information, the openness of information to other people, the appropriateness of certain indicators to specific networks, and whether the indicator configuration is all-encompassing, and equipped for uncovering the propensities of cultural capacities. These are a portion of the perplexing issues that despite everything need settling through more prominent global coordination and participation. Since the structure of pointers depends on esteem suspicions, markers created to quantify feasible improvement are ideally a multi-partner activity; as needs are, a pluralistic viewpoint of what establishes sustainability will be spoken to in the design (Fox et al., 2013; Grzebyk and Stec, 2015).

A long term checking system to gather information on key parts of the worldwide nature and the human network will give multidisciplinary knowledge about the world. The estimations of cultural advancement and worldwide biology will add to the making of a worldwide database, and this will fill in as an important scholarly supply for directors, organizers, and chiefs at all levels. The Global Environmental Outlook (GEO), a generally new UNEP production, which gives a state-of-the-art evaluation of the condition of the worldwide condition and policy reactions, and other such distributions are subordinates of this sort of checking and information assortment process (Mashishi, 2001; North American Commission for Environmental Co-activity, 2002). Thirty-five GEO bases have been built upon the world to gather and blend information of territorial and worldwide value. Numerous countries have additionally started to distribute the condition of the earth reports as a type of public data. Huge partnerships have additionally been giving sustainability execution audits of their tasks not just for their investors, however more critically for their stakeholders. Stakeholder's commitment is viewed as a fundamental segment of a sustainable endeavor business approach. So also, numerous NGOs distribute their information and viewpoints for their participation and the overall population. Supporting the numerous pages of words and figures depicting and examining ecological conditions, and the advancement that has been made in propelling the sustainable improvement cause, there are various research forms at work producing data at each level and in each division of society (Zhao et al., 1999; Tripathi et al., 2017).

Detailing is frequently utilized as an observing and specialized device required by multilateral treaties. Under the Framework Convention on Climate Change, all parties must report at standard interims to the meeting of the gatherings. The underlying report must incorporate the national stock of anthropogenic greenhouse gas emanations information, joined by an execution plan and important supporting data (Matsuda et al., 2012; Lamidi et al., 2019).

New information technologies (IT) have made exceptional open doors for sharing and trading data. The World Wide Web, for instance, has opened new roads for open interest in a wide range of worldwide issues. Here and there, the data innovations have incredibly encouraged openness of data for general society. They empower the instantaneous sharing of data by various clients and, along these lines; numerous new sorts of global participation can be possible. The World Wide Web can convey satellite-based information service to any Internet user (see "Maintainability, information the board, and the Internet").

There has been a lot of accentuation on scientific solutions to help sustainable advancement. In the Convention on the Conservation of Biological Diversity (1992), the United Nations Framework Convention on Climate Change (1992) and later the Agreement on Desertification (1994) the logical and technological capacity to assess natural execution, distinguish issues, and discover arrangements have been underlined. Numerous lawful treaties additionally make solutions for the making of scientific and technological bodies. As an ongoing United Nations report called attention to, be that as it may, there are wide holes between distinguishing requirements for solutions and the real fulfillment of viable solutions for our environmental issues. There is a requirement for a more noteworthy coordinated effort among the logical trains in developing measures for accomplishing sustainable advancement (Lago et al., 2010; Doula et al., 2017).

Sadly, notwithstanding the issue of deficient advancement in science and innovation look into, innovation move to the third world is additionally hampered by an absence of help with subsidizing. This is identified with the worldwide phenomenon of public sector offloading. Lately, the private sector and the market system, much helped by a progression of changes towards more prominent advancement of investment and international trade, are dared to have taken on a greater role in conveying the government assistance work locally and universally. An ongoing report of the Commission on SD called attention to that since the Earth Summit, the volume of overseas development aid (ODA) has been on a general descending pattern, from US$58.3 billion of every 1992 to US$53.1 billion of every 2000. The appropriation of help, during this period, has likewise been rebuilt by the donor nations as indicated by their cost-effectiveness evaluations. Subsequently, the majority of the least evolved nations endured a decrease in help of in any event 25%. This decrease in ODA, be that as it may, has been joined by an expansion in foreign direct investment right now. The expansion in foreign direct investment has been lopsided; just ten nations got around 80% of the aggregate sum. Simultaneously, consumption on arms and the military has been expanding, especially after September 11, 2001, and President George Bush declared in his 2002 State of the Union address that there would be a further increment to the US$1 billion every month military spending on the present war. In a matter of seconds a short time later Bush's Security Council referenced the extra request was US$48 billion. This expansion alone would have gone far towards fulfilling a wide scope of worldwide needs identified with sustainable improvement (Harvey, 2013; Obersteiner et al., 2016).

The decrease in accessibility of budgetary aid and the reassessment of subsidizing criteria have not been useful for some developing nations, which were sick prepared in any case to adapt to meeting conventional essential needs. The absence of monetary help to back the responsibilities made at the Earth Summit has been accused of the moderate improvement in Agenda 21 initiatives in underdeveloped nations and somewhere else (Long et al., 2004; Vadoudi et al., 2014).

Unmistakably the private segment, which creates the vast majority of the world's wealth, has a key job in economical advancement in both the North and the South. The WBCSD paper illustrating "The Business Case for SD" presents convincing contentions for the selection of sustainability as a key imperative for the twenty-first century. The United Nations has presented the "Global Compact" to connect with the private sector in multi-partner gatherings, to embrace worldwide citizenship standards, and to take an interest in the sustainable advancement process. Regardless of the way that bold steps have been taken in building new establishments, confining new dynamic procedures, making new lawful systems, encouraging exploration systems, and so forth., nobody can yet guarantee ownership of an amazing plan for sustainability. Without this conclusive outline, all associations setting out on the way toward sustainability must utilize the standards of "adaptive management" given consistent learning and reaction to the feedback of different sorts.

The best challenge to the usage of sustainable improvement lies in the definition of a joint learning, co-developing procedure, as examined in "Culture and sustainability." The trouble begins in the ill adjustment of human culture to nature. Sustainable improvement is tied in with discovering answers for issues that are characteristic in the human system. Humankind's capacity to identify, analyze, and recommend answers for those issues is constrained by our degree of awareness, the limit of our institutions to change our general public, and the capacity of our general public to accomplish consensus, and to look for solutions. Sciences offer one part of human innovativeness. Research structure for economical improvement requires the coordination of both technical studies, and arts of the human experience. To date, the one most significant fixing missing from the formula for accomplishing sustainability is political will. Developing sustainability champions and pioneers at all levels from the nearby to the global and overall divisions of society has gotten an earnest goal (Peng et al., 2013).

REFERENCES

1. Altieri, M. A., Funes-Monzote, F. R., & Petersen, P., (2012). Agroecologically efficient agricultural systems for smallholder farmers: Contributions to food sovereignty. *Agronomy for Sustainable Development*, *32*(1), 1–13.
2. Azapagic, A., (2004). Developing a framework for sustainable development indicators for the mining and minerals industry. *Journal of Cleaner Production*, *12*(6), 639–662.
3. Bansal, P., (2005). Evolving sustainably: A longitudinal study of corporate sustainable development. *Strategic Management Journal*, *26*(3), 197–218.
4. Bell, D. V. J., & Myers, J., (2001). *Seizing the Opportunities, Seizing the Future: A Discussion Paper for Natural Resources Canada* (Vol. 1, p. 55). http://www.yorku.ca/ycas/publications/index.html (accessed on 1 May 2020).
5. Bell, D. V. J., & Schwartzberg, P., (2001). *Advancing Community Sustainable Development in the Ontario Region: The Role of the Government of Canada* (Vol. 1, pp. 20–27). http://www.yorku.ca/ycas/publications/index.html (accessed on 1 May 2020).
6. Bell, D., Halucha, P., & Hopkins, M., (2001). *Sustainable Development Concept Paper*, *1*, pp. 10–19.
7. Berno, T., & Bricker, K., (2001). Sustainable tourism development: The long road from theory to practice. *International Journal of Economic Development*, *3*(3), 1–18.
8. Bossel, H., (1999). *Indicators for Sustainable Development: Theory, Method, Applications* (Vol. 1, p. 138). Winnipeg: International Institute for Sustainable Development.
9. Brumano, L. P., Soler, M. F., & Da Silva, S. S., (2016). Recent advances in sustainable production and application of biosurfactants in Brazil and Latin America. *Industrial Biotechnology*, *12*(1), 31–39.
10. Cash, D. W., Clark, W. C., Alcock, F., Dickson, N. M., Eckley, N., Guston, D. H., & Mitchell, R. B., (2003). Knowledge systems for sustainable development. *Proceedings of the National Academy of Sciences*, *100*(14), 8086–8091.
11. Cheung, Y.-K. A., (1997). Chapter two: The concept of sustainable development. An environmental assessment approach for Hong Kong

and the Pearl River delta: Principles and practices. *PhD Thesis* (Vol. 1, pp. 40–49). The University of Hong Kong.

12. Cobbinah, P. B., Erdiaw-Kwasie, M. O., & Amoateng, P., (2015). Africa's urbanization: Implications for sustainable development. *Cities*, *47*, 62–72.
13. Common, M., & Perrings, C., (1992). Towards an ecological economics of sustainability. *Ecological Economics*, *6*(1), 7–34.
14. Costanza, R., & Daly, H. E., (1992). Natural capital and sustainable development. *Conservation Biology*, *6*(1), 37–46.
15. D'souza, G., & Ikerd, J., (1996). Small farms and sustainable development: Is small more sustainable? *Journal of Agricultural and Applied Economics*, *28*(1), 73–83.
16. Davis, K., Nkonya, E., Kato, E., Mekonnen, D. A., Odendo, M., Miiro, R., & Nkuba, J., (2012). Impact of farmer field schools on agricultural productivity and poverty in East Africa. *World Development*, *40*(2), 402–413.
17. De Bon, H., Parrot, L., & Moustier, P., (2010). Sustainable urban agriculture in developing countries. A review. *Agronomy for Sustainable Development*, *30*(1), 21–32.
18. Development (Rio de Janeiro, (1992)). (Vol. 1, pp. 1–5.). http://www.un.org/documents/ga/conf151/aconf15126-1annex1.htm (accessed on 1 May 2020).
19. Dincer, I., (2000). Renewable energy and sustainable development: A crucial review. *Renewable and Sustainable Energy Reviews*, *4*(2), 157–175.
20. Dogliotti, S., Van Ittersum, M. K., & Rossing, W. A. H., (2005). A method for exploring sustainable development options at farm scale: A case study for vegetable farms in South Uruguay. *Agricultural Systems*, *86*(1), 29–51.
21. Dolzer, R., & Thesing, J., (2000). *Protecting Our Environment: German Perspectives on a Global Challenge* (Vol. 1, pp. 570–590). Germany: Konrad-Adenauer-Stiftung.
22. Doula, M. K., Moreno-Ortego, J. L., Tinivella, F., Inglezakis, V. J., Sarris, A., & Komnitsas, K., (2017). Olive mill waste: Recent advances for the sustainable development of olive oil industry. In: *Olive Mill Waste* (Vol. 1, pp. 29–56). Academic Press.

23. Eden, S. E., (1994). Using sustainable development: The business case. *Global Environmental Change*, *4*(2), 160–167.
24. Ekpo, E. I., (2012). Impact of shipping on nigerian economy: Implications for sustainable development. *Journal of Educational and Social Research*, *2*(7), 107–117.
25. Elkington, J., (1994). Towards the sustainable corporation: Win-win-win business strategies for sustainable development. *California Management Review*, *36*(2), 90–100.
26. Ferrer-Balas, D., Bruno, J., De Mingo, M., & Sans, R., (2004). Advances in education transformation towards sustainable development at the Technical University of Catalonia, Barcelona. *International Journal of Sustainability in Higher Education, 1*, 1–20.
27. *Foods Called Beneficial and Safe.* (Vol. 1, pp. 1–10). http://www.plant.uoguelph.ca/riskcomm/archives/agnet/(2000)/1–(2000)/1g-01–23–00–01.txt (accessed on 1 May 2020).
28. Foster, J., (2001). *Knowing Ourselves: A Brief History of Emerging Global Civil Society: Prepared for the 4th CIVICUS World Assembly August (2001)* (Vol. 1, pp. 5–10). Vancouver, B.C., Canada.
29. Fox, E. B., Liu, Z. W., & Liu, Z. T., (2013). Ultraclean fuels production and utilization for the twenty-first century: Advances toward sustainable transportation fuels. *Energy and Fuels*, *27*(11), 6335–6338.
30. Gao, W. W., Zhao, Y. J., Wang, Y. P., & Chen, S. L., (2006). A review of research on sustainable use of medicinal plants cropland in China. *China Journal of Chinese Materia Medica*, *31*(20), 1665–1669.
31. Gherasim, M. E., & Tanase, G., (2012). The fundamentals of sustainable development. *Contemporary Readings in Law and Social Justice*, *4*(1), 446.
32. Gibbs, D., (1994). The implications of sustainable development for industry and employment in the 1990s. *Environmentalist*, *14*(3), 183–192.
33. Glémain, P., (2011). The strategy and fundamentals of sustainable finance serving sustainable development. *Finance and Sustainability: Towards a New Paradigm*, *1*, 187–209.
34. Griggs, D., Stafford-Smith, M., Gaffney, O., Rockström, J., Öhman, M. C., Shyamsundar, P., & Noble, I., (2013). Policy: Sustainable development goals for people and planet. *Nature*, *495*(7441), 305.

35. Grzebyk, M., & Stec, M., (2015). Sustainable development in EU countries: Concept and rating of levels of development. *Sustainable Development*, *23*(2), 110–123.

36. Gupta, J., & Vegelin, C., (2016). Sustainable development goals and inclusive development. *International Environmental Agreements: Politics, Law and Economics*, *16*(3), 433–448.

37. Hák, T., Janoušková, S., & Moldan, B., (2016). Sustainable development goals: A need for relevant indicators. *Ecological Indicators*, *60*, 565–573.

38. Harvey, L. D., (2013). Recent advances in sustainable buildings: Review of the energy and cost performance of the state-of-the-art best practices from around the world. *Annual Review of Environment and Resources*, *38*, 281–309.

39. Holt, R. P., (2005). Post-Keynesian economics and sustainable development. *International Journal of Environment, Workplace and Employment*, *1*(2), 174–186.

40. Holthus, P., & Council, M. A., (1999). Sustainable development of oceans and coasts: The role of the private sector. In: *Natural Resources Forum* (Vol. 23, No. 2, pp. 169–176). London: Butterworths.

41. Hörisch, J., (2016). Entrepreneurship as facilitator for sustainable development? *Editorial for the Special Issue: Advances in Sustainable Entrepreneurship, 1*, 1–19.

42. Intergovernmental Panel on Climate Change, (2001). *Summary for Policymaker, a Report of Working Group 1 of the Intergovernmental Panel on Climate Change* (Vol. 1, pp. 10–20). http://www.ipcc.ch (accessed on 1 May 2020).

43. International Council for Local Environmental Initiatives (ICLEI)., (1999). CAG consultants and the division for sustainable development of UNDESA. *Study on National Obstacles to Local Agenda 21* (Vol. 1, pp. 20–26). http://www.un.org/esa/sustdev/iclei.htm (accessed on 1 May 2020).

44. Jolink, A., & Niesten, E., (2015). Sustainable development and business models of entrepreneurs in the organic food industry. *Business Strategy and the Environment*, *24*(6), 386–401.

45. Jordan, N., Boody, G., Broussard, W., Glover, J. D., Keeney, D., McCown, B. H., & Pansing, C., (2007). Sustainable development of the agricultural bio-economy. *Science*, *316*(5831), 1570–1571.

46. Juana, J. S., Kahaka, Z., & Okurut, F. N., (2013). Farmers' perceptions and adaptations to climate change in sub-Sahara Africa: A synthesis of empirical studies and implications for public policy in African agriculture. *Journal of Agricultural Science*, *5*(4), 121.
47. Lago, L. I., Ponta, F. L., & Chen, L., (2010). Advances and trends in hydrokinetic turbine systems. *Energy for Sustainable Development*, *14*(4), 287–296.
48. Lamidi, R. O., Jiang, L., Pathare, P. B., Wang, Y., & Roskilly, A. P., (2019). Recent advances in sustainable drying of agricultural produce: A review. *Applied Energy*, *233*, 367–385.
49. Lawn, P. A., (2001). Goods and services and the dematerialization fallacy: implications for sustainable development indicators and policy. *International Journal of Services Technology and Management*, *2*(3/4), 363–376.
50. Le Blanc, D., (2015). Towards integration at last? The sustainable development goals as a network of targets. *Sustainable Development*, *23*(3), 176–187.
51. Lélé, S. M., (1991). Sustainable development: A critical review. *World Development*, *19*(6), 607–621.
52. Lerner, S., (1992). *Environmental Stewardship: Studies in Active Earth Keeping* (Vol. 1, pp. 445–453). Department of Geography Publication Series, No.39, University of Waterloo, Canada.
53. Lichtfouse, E., Navarrete, M., Debaeke, P., Souchère, V., Alberola, C., & Ménassieu, J., (2009). Agronomy for sustainable agriculture: A review. In: *Sustainable Agriculture* (Vol. 1, pp. 1–7). Springer, Dordrecht.
54. Long, T. R., Jiang, W. C., & He, Q., (2004). Water resources carrying capacity: New perspectives based on eco-economic analysis and sustainable development. *Journal of Hydraulic Engineering*, *1*, 38–45.
55. Ma, Y., Chen, L., Zhao, X., Zheng, H., & Lü, Y., (2009). What motivates farmers to participate in sustainable agriculture? Evidence and policy implications. *International Journal of Sustainable Development and World Ecology*, *16*(6), 374–380.
56. Martin, B., McGutre, F., & Allen, L., (1998). Retirees' attitudes toward tourism: Implications for sustainable development. *Tourism Analysis*, *3*(1), 43–51.

57. Martinelli, L. A., Naylor, R., Vitousek, P. M., & Moutinho, P., (2010). Agriculture in Brazil: Impacts, costs, and opportunities for a sustainable future. *Current Opinion in Environmental Sustainability*, *2*(5/6), 431–438.

58. Mashishi, M., (2001). *Time for a New Global Compact, Daily Mail and The Guardian* (Vol. 1, pp.1–10). UK. http://www.mg.co.za/mg/archive/(2001)oct/features/15octcompact.html (accessed on 1 May 2020).

59. Matsuda, K., Iwakabe, K., & Nakaiwa, M., (2012). Recent advances in internally heat-integrated distillation columns (HIDiC) for sustainable development. *Journal of Chemical Engineering of Japan*, *45*(6), 363–372.

60. Mebratu, D., (1998). Sustainability and sustainable development: Historical and conceptual review. *Environmental Impact Assessment Review*, *18*(6), 493–520.

61. Meister, M., & Japp, P. M., (1998). Sustainable development and the global economy: Rhetorical implications for improving the quality of life. *Communication Research*, *25*(4), 399–421.

62. Milanez, B., & De Oliveira, J. A. P., (2013). Innovation for sustainable development in artisanal mining: Advances in a cluster of opal mining in Brazil. *Resources Policy*, *38*(4), 427–434.

63. Mitchell, G., (1996). Problems and fundamentals of sustainable development indicators. *Sustainable Development*, *4*(1), 1–11.

64. Mitchell, G., (2008). *Problems and Fundamentals of Sustainable Development Indicators*, *1*, pp. 1–30.

65. Montana, J. P., & Nenide, B., (2008). The evolution of regional industry clusters and their implications for sustainable economic development: Two case illustrations. *Economic Development Quarterly*, *22*(4), 290–302.

66. Mutekwa, V. T., (2009). Climate change impacts and adaptation in the agricultural sector: The case of smallholder farmers in Zimbabwe. *Journal of Sustainable Development in Africa*, *11*(2), 237–256.

67. Muzari, W., Gatsi, W., & Muvhunzi, S., (2012). The impacts of technology adoption on smallholder agricultural productivity in sub-Saharan Africa: A review. *Journal of Sustainable Development*, *5*(8), 69.

68. Nilsson, M., Griggs, D., & Visbeck, M., (2016). Policy: Map the interactions between sustainable development goals. *Nature*, *534*(7607), 320–322.

69. North American Commission for Environmental Co-operation, (2002). *The North American. Mosaic: The State of the Environment Report* (Vol. 1(11), pp. 1-3). http://www.cec.org/about-us/jpac-advice-council/advice-council-02-09.

70. Obersteiner, M., Walsh, B., Frank, S., Havlík, P., Cantele, M., Liu, J., & Valin, H., (2016). Assessing the land resource-food price nexus of the sustainable development goals. *Science Advances*, *2*(9), e1501499.

71. Omer, A. M., (2008). Energy, environment, and sustainable development. *Renewable and Sustainable Energy Reviews*, *12*(9), 2265–2300.

72. Orr, D. W., (1991). *Ecological Literacy: Education and the Transition to a Postmodern World* (Vol. 1, pp. 200–210). Albany, State University of New York Press.

73. Orsini, F., Kahane, R., Nono-Womdim, R., & Gianquinto, G., (2013). Urban agriculture in the developing world: A review. *Agronomy for Sustainable Development*, *33*(4), 695–720.

74. Parris, T. M., & Kates, R. W., (2003). Characterizing and measuring sustainable development. *Annual Review of Environment and Resources*, *28*(1), 559–586.

75. Paul, B. G., & Vogl, C. R., (2011). Impacts of shrimp farming in Bangladesh: Challenges and alternatives. *Ocean and Coastal Management*, *54*(3), 201–211.

76. Peng, Q., Liu, Y., Gu, P., & Fan, Z., (2013). Development of an open-architecture electric vehicle using adaptable design. In: *Advances in Sustainable and Competitive Manufacturing Systems* (Vol. 1, pp. 79–90). Springer, Heidelberg.

77. Prakash, C. S., (2000). *Scientists Release Declaration Defending Biotechnology: Genetically Modified*. PR Newswire, Institute for Agriculture and Trade Policy (Vol.1, pp. 1-2).

78. Pretty, J. N., (1995). Participatory learning for sustainable agriculture. *World Development*, *23*(8), 1247–1263.

79. Prud'homme, B., & Raymond, L., (2013). Sustainable development practices in the hospitality industry: An empirical study of their impact on customer satisfaction and intentions. *International Journal of*

Hospitality Management, *34*, 116–126.

80. Rigby, D., Woodhouse, P., Young, T., & Burton, M., (2001). Constructing a farm level indicator of sustainable agricultural practice. *Ecological Economics*, *39*(3), 463–478.
81. Righi, E., Dogliotti, S., Stefanini, F. M., & Pacini, G. C., (2011). Capturing farm diversity at regional level to up-scale farm level impact assessment of sustainable development options. *Agriculture, Ecosystems and Environment*, *142*(1/2), 63–74.
82. Robert, K. W., Parris, T. M., & Leiserowitz, A. A., (2005). What is sustainable development? Goals, indicators, values, and practice. *Environment: Science and Policy for Sustainable Development*, *47*(3), 8–21.
83. Röling, N., & Wagemakers, M. A. E., (1998). *Facilitating Sustainable Agriculture: Participatory Learning and Adaptive Management in Times of Environmental Uncertainty*, *1*, 1–20.
84. Rosen, M. A., (2019). Advances in sustainable development research. *European Journal of Sustainable Development Research*, *3*(2), 1–20.
85. Sachs, J. D., (2012). From millennium development goals to sustainable development goals. *The Lancet*, *379*(9832), 2206–2211.
86. Sadorsky, P., (2014). The effect of urbanization and industrialization on energy use in emerging economies: Implications for sustainable development. *American Journal of Economics and Sociology*, *73*(2), 392–409.
87. Schaper, M., (2002). The challenge of environmental responsibility and sustainable development: Implications for SME and entrepreneurship academics. *Radical Changes in the World: Will SMEs Soar or Crash*, *1*, 541–553.
88. Schindler, J., Graef, F., & König, H. J., (2015). Methods to assess farming sustainability in developing countries: A review. *Agronomy for Sustainable Development*, *35*(3), 1043–1057.
89. Scholz, M., Hedmark, Å., & Hartley, W., (2012). Recent advances in sustainable multifunctional land and urban management in Europe: A review. *Journal of Environmental Planning and Management*, *55*(7), 833–854.
90. Scopel, E., Triomphe, B., Affholder, F., Da Silva, F. A. M., Corbeels, M., Xavier, J. H. V., & De Carvalho Mendes, I., (2013). Conservation agriculture cropping systems in temperate and tropical conditions,

performances, and impacts: A review. *Agronomy for Sustainable Development*, *33*(1), 113–130.

91. Sharpley, R., (2000). Tourism and sustainable development: Exploring the theoretical divide. *Journal of Sustainable Tourism*, *8*(1), 1–19.
92. Shiyi, C., (2009). Energy consumption, CO_2 emission, and sustainable development in Chinese industry [J]. *Economic Research Journal*, *4*, 1–5.
93. Skerlos, S. J., Hayes, K. F., Clarens, A. F., & Zhao, F., (2008). Current advances in sustainable metalworking fluids research. *International Journal of Sustainable Manufacturing*, *1*(1/2), 180–202.
94. Smith, C. S., & McDonald, G. T., (1998). Assessing the sustainability of agriculture at the planning stage. *Journal of Environmental Management*, *52*(1), 15–37.
95. Smith, S. L., (1995). Ecologically sustainable development: Integrating economics, ecology, and law. *Willamette L. Rev.*, *31*, 261.
96. Solow, R. M., (1986). On the intergenerational allocation of natural resources. *The Scandinavian Journal of Economics*, *1*, 141–149.
97. Stafford-Smith, M., Griggs, D., Gaffney, O., Ullah, F., Reyers, B., Kanie, N., & O'Connell, D., (2017). Integration: The key to implementing the sustainable development goals. *Sustainability Science*, *12*(6), 911–919.
98. Stave, J. W., & Durandetta, D., (2000). GM crop testing grows amid controversy. *Today's Chemist at Work*, *1*, *9*(6), 1–9, 32, 33, 37. http://pubs.acs.org/hotartcl/tcaw/00/jun/stave.html (accessed on 1 May 2020).
99. Timur, S., & Getz, D., (2009). Sustainable tourism development: How do destination stakeholders perceive sustainable urban tourism? *Sustainable Development*, *17*(4), 220–232.
100. Tripathi, V., Edrisi, S. A., Chen, B., Gupta, V. K., Vilu, R., Gathergood, N., & Abhilash, P. C., (2017). Biotechnological advances for restoring degraded land for sustainable development. *Trends in Biotechnology*, *35*(9), 847–859.
101. Tueth, M., (2009). Fundamentals of sustainable business: A guide for the next 100 years. *World Scientific Books*, *1*, pp. 1–30.
102. UNCTAD, (2000a). *Management of Biotechnology Seen as Crucial to Developing World* (Vol. 1, pp. 1–10). https://www.un.org/press/en/2000/20000719.tad1920.doc.html.
103. UNCTAD, (2000b). *We Can't Turn Back the Clock on Globalization,*

World Bank President Tells UN Trade and Development Conference (Vol. 1, pp. 1–11). https://www.un.org/press/en/2000/20000216.tad1913.doc.html.

104. United Nations Economic and Social Council, (2001). *Implementing Agenda 21: Report of the Secretary General, 1*, pp. 50–63.

105. United Nations General Assembly, (2000). *Report of the United Nations Conference on Environment and Development, Vol.1, pp. 1-13.*

106. United Nations Population Fund, (2001). *Chapter 2: Environmental Trends* (Vol. 1, pp. 10–21). The State of World Population. http://www.unfpa.org/swp/(2001)/english/ch02html (accessed on 1 May 2020).

107. United Nations Press Release, (2000a). *Commission on Sustainable Development Concludes Two-Day Dialogue on Sustainable Agriculture* (Vol. 1, pp. 1–10). https://www.un.org/press/en/2000/20000425.envdev541.doc.html.

108. United Nations Press Release, (2000b). *Commission on Sustainable Development Holds Dialogue on Knowledge for a Sustainable Food System* (Vol. 1, pp. 1–10). https://sustainabledevelopment.un.org/index.php?menu=1418.

109. US Office of the Press Secretary, (2000a). *President Delivers State of the Union Address* (Vol. 1, pp. 1–9). http://www.whitehouse.gov/news/releases/2002/01/20020129–11.html (accessed on 1 May 2020).

110. US OFFICE of the Press Secretary, (2000b). *National Security Advisor Speaks at Conference, Remarks by the National Security Advisor Condoleezza Rice to the Conservative Political Action Conference* (Vol. 1, pp. 1–6).

111. Vadoudi, K., Allais, R., Reyes, T., & Troussier, N., (2014). Sustainable product lifecycle management and territoriality: New structure for PLM. In: *IFIP International Conference on Product Lifecycle Management* (Vol. 1, pp. 475–484). Springer, Berlin, Heidelberg.

112. Van de Fliert, E., & Braun, A. R., (2002). Conceptualizing integrative, farmer participatory research for sustainable agriculture: From opportunities to impact. *Agriculture and Human Values, 19*(1), 25–38.

113. Van der Werf, H. M., & Petit, J., (2002). Evaluation of the environmental impact of agriculture at the farm level: A comparison and analysis of 12 indicator-based methods. *Agriculture, Ecosystems and Environment, 93*(1–3), 131–145.

114. Wackernagel, M., & Rees, W., (1996). *Our Ecological Foot-Print:*

Reducing Human Impact on the Earth (Vol. 1, pp. 150–160). Gabriola Island, B.C. and Philadelphia, PA, New Society Publishers.

115. WBCSD and UNDEP, (1999) *Eco-Efficiency and Cleaner Production: Charting the Course to Sustainability* (Vol. 1, pp. 1–13). http://www.iisd.ca/linkages/consume/unep.html (accessed on 1 May 2020).
116. WCED, (1987). *Our Common Future: World Commission on Environment and Development* (Vol. 1, pp. 390–400). New York, Oxford University Press.
117. Welford, R., (2013). Models of sustainable development for business. In: *Hijacking Environmentalism* (Vol. 1, pp. 179–210). Routledge.
118. World Trade Organization, (2001). *Statistics on Globalization* (Vol. 1, pp. 40–44). http://www.wto.org (accessed on 1 May 2020).
119. Wu, T. Y., Mohammad, A. W., Lim, S. L., Lim, P. N., & Hay, J. X. W., (2013). Recent advances in the reuse of wastewaters for promoting sustainable development. In: *Wastewater Reuse and Management* (Vol. 1, pp. 47–103). Springer, Dordrecht.
120. Zahm, F., Viaux, P., Vilain, L., Girardin, P., & Mouchet, C., (2008). Assessing farm sustainability with the IDEA method-from the concept of agriculture sustainability to case studies on farms. *Sustainable Development*, *16*(4), 271–281.
121. Zander, P., & Kächele, H., (1999). Modeling multiple objectives of land use for sustainable development. *Agricultural Systems*, *59*(3), 311–325.
122. Zhao, J., Liang, X., & Zhang, X., (1999). System analysis on the definition of sustainable development. *Acta Ecologica Sinica*, *19*(3), 393–398.

CHAPTER 2

Sustainable Development and Environment

CONTENTS

2.1. Introduction 40
2.2. Environmental Resources 40
2.3. Environmental Indicators 41
2.4. Environment and Sustainable Development (SD) 42
2.5. Environmental Interdependence 45
2.6. Problems Related to Environment And International Trade 46
2.7. Environment and Economic Growth, Trade-Off 47
2.8. Externalities Impact on Environment 48
2.9. Environmental Models 49
2.10. Immoderate Utilization of Resources 54
2.11. Developed Countries (DCS) Contribution in Environmental Pollution 55
2.12. Environmental Hazards Affecting Human Health 56
2.13. Environment and Market Incentives 58
References 60

2.1. INTRODUCTION

In early 1970, the concept of the environment started evolving as it became a worldwide issue. Initially, it was acknowledged worldwide that the ecosystem of the Earth is fragile, and human beings have been significantly contributing to its degradation (World Bank, 2003). Several countries began their mutual efforts in order to strike a balance amid safeguarding the environment and improving the standard of human life for future generations which gave rise to a new awareness (John and Pecchenino, 1997; Holgate et al., 1999). The economic, as well as the social welfare of humankind, is closely associated with the environment. Any type of change whether positive or negative in the socio-economic field will immediately or eventually cause an impact on the environment of the Earth. In numerous such cases, the results of negative impacts cannot be reversed (Laffont and Tirole, 1994; Shafik, 1994; Selden and Song, 1995). In 1992, Earth Summit took place in Rio in which it was concluded that social, economic, as well as environmental issues are undeniably interlinked with the development of the world. Therefore, it was pledged in the summit that we should eliminate environmental problems, decrease poverty, and achieve sustainable development (SD) by means of joint efforts and international cooperation (Hettiqe, 1997; Haq, 1999; Barrett and Graddy, 2000).

2.2. ENVIRONMENTAL RESOURCES

Those resources which have innate values and are existing naturally for the continuous usability of humans are termed as environmental resources. Referring to economic terms, huge extents of environmental resources are not tangible (Tennant, 1976; Mäler, 1991; Gregory et al., 1993). Human and social resources are described depending upon the population group's denseness, their professional affairs, rights they pose on their land, the quality of life they live, and other aspects concerned with gender, etc. The rate at which global resources are being used is surely one of the crucial universal problems. Population increase results in more need of resources to counter the needs of the upcoming generation. We put light on four categories of environmental, human, and natural resources in combined terms (Benotsch et al., 2000; Stevens and Olsen, 2003). Continuous, extrinsic, non-renewable, and renewable are four categories of resources (Bromet and Moos, 1977).

2.2.1. Renewable Resources

Within a moderate time span, renewable resources possess the capability of naturally regenerating themselves in useful products. They are competent enough not only in renewing themselves but also in making themselves available until natural disaster breaks in and ends them. Fauna, flora, clean air, and clean water are prominent examples of renewable resources.

2.2.2. Non-Renewable Resources

A limited amount of non-renewable resources are available. We can say that the extent to which these resources are present is restricted and this is due to their slow renewable rate. Minerals, groundwater, etc are examples of such resources.

2.2.3. Continuous Resources

Even the gross mismanagement will not be able to degrade the availability of continuous resources. Geothermal and tidal energy, gravity, wind, and solar energy are examples of such resources. Solar energy might be an exceptional case as it is affected by pollution.

2.2.4. Extrinsic Resources

Debased and devalued resources, but if managed appropriately, they are always available such as institution management abilities, human skills, etc.

2.3. ENVIRONMENTAL INDICATORS

Environmental, economic, and social indicators, all are linked with each other to an extent that we are able to recognize the level that how much developed a country is, though few specific environmental indicators, which are followings (Ang et al., 1998; Olsthoorn et al., 2001; Bockstaller and Girardin, 2003; Piorr, 2003):

- Minimal quantity of waste.
- Designs of environments that are based on nature.
- The proportion of potable water used by public.
- The proportion of sewers being used by the public.
- The proportion of the public who are using carpooling or public transport.

- The proportion of agricultural land.
- The proportion of recycling programs in which households are involved.
- Amount of generated hazardous waste on a yearly bases.
- Amount of generated solid waste per capita on yearly bases.
- Amount of yearly toxic release.
- The number of air days which were clean.
- Depending upon the number of institutes who are addressing environmental issues.
- Depending upon the number of academic institutes whether public or private, which are addressing environmental education.
- Legal environmental structure.
- The extent to which people are aware of the benefits provided by a clean environment.
- The extent to which the public is sensitive to pollution.
- Allotment of the amount fixed in the budget for environmental projects.
- Percentage of chemical fertilizers and chemicals being used.
- Levels of hazards one might face in his physical health state.
- A framework managing the nature of waste.

2.4. ENVIRONMENT AND SUSTAINABLE DEVELOPMENT (SD)

There exists a close relationship between the SD and the environment. Leaving distinction behind, in a broader view, this relationship also extends to the complete development of mankind. International Union for Conservation of Nature (IUCN), and the United Nations Environment Program (UNEP) introduced the World Conservative Strategy which serves as a platform for global discussions on sustainability. In 1987, the World Commission on Environment and Development (WCED) published a report which was titled "Our Common Future" and mostly called "The Brundtland Report." This report has remarkable significance in SD. SD was defined in this report as "Development that meets the needs of the present, without compromising the ability of future generations to meet their own needs"

(Azapagic, 2004; Altieri et al., 2012).

As stated by the report, the utmost goal of development must be to satisfy human requirements and material desires. It focuses on the reality that misuse of resources may lead to a point where human societies would have to compromise their capability to cope up with people's basic requirements in the future. This report mentions human intervention examples as genetic modification, commercial use of forests, heat emission, emission of toxic gases, mineral extraction, watercourse diversion, and settled agriculture. It was suggested to all countries to acquire and include SD as the primary goal of their national policy and international cooperation (Azapagic, 2004; Bansal, 2005).

In 1992, 1997, and 2002, the United Nations Conference on Environment and Development (UNCED) held three different Earth Summits in which more than 170 countries took part. All of the countries revived their dedication to SD targeting the objective at "worldly conditions that impose threats to people may be given extra attention including corruption, natural disasters, drug problems, long-term hunger, disputes among armies, malnutrition, transferrable diseases such as HIV/AIDS, tuberculosis, and malaria" (Stave and Durandetta, 2000; Bansal, 2005; Hopwood et al., 2005).

The union of human wisdom and opportunities provided by the environment now serves as the universally applicable theme for SD which became possible only due to these summits (Timur and Getz, 2009; Tripathi et al., 2017).

2.4.1. Aspects of Sustainable Development (SD)

There are many dimensions of SD. Following are a few of them discussed in detail (Schaper, 2002; Scholz et al., 2012; Schindler et al., 2015):

2.4.1.1. Social Dimension

It comprises of (Prakash, 2000; Giddings et al., 2002; Peng et al., 2013):

- Safety and health of workers;
- Standard of living, the effect on local communities;
- The benefits are given to disadvantaged groups. Disabled ones are an example.

2.4.1.2. Economic Dimension

It comprises of (Lawn, 2001; Holt, 2005; Harvey, 2013):

- Chances of sale growth and new markets created;
- Additional values being created;
- Reducing the cost by utilizing less energy and fewer raw materials and by improving efficiency.

2.4.1.3. Environmental Dimension

It comprises of (Omer, 2008; Martinelli et al., 2010; Lamidi et al., 2019):

- Reducing the discharge of waste in the environment and by lessening the sewage;
- Reducing the influence on humans;
- Utilizing raw materials which are viable;
- By discarding the use of substances that are toxic.

With the objective of improving, for the betterment of human beings, development possesses a different nature. Given below is the outline of these different natures that development possesses (Pretty, 1995; Muzari et al., 2012:

- Human Development: Social structure and other social aspects are discussed here.
- Cultural Development: The lifestyle, cultural norms, and traditions being followed are discussed here.
- Spiritual Development: The righteous, believing, and noble values are discussed here.
- Ecological Development: Environment is discussed here. We talk about nature, its concerns, and requirement.
- Economic Development: Economic system and Economy are discussed here.
- Political Development: System of politics and rules of government are discussed here.

2.4.2. Attainment of Sustainable Development (SD)

Mandatory needs are required to be fulfilled to reach out to the basic aim of SD. Empowering under-listed sections of society will help in achieving this aim (Martin et al., 1998; Ma et al., 2009; Nilsson et al., 2016).

- Women;
- Youngsters;

- Children;
- Farmers;
- Local communities and their native peoples;
- Non-governmental organization NGOs;
- Municipality, i.e., giving authority to town government;
- Trade unions and their workers;
- Industries and businesses;
- An organization that is technological or scientific.

2.5. ENVIRONMENTAL INTERDEPENDENCE

As emphasized by the Environmentalists, sustainable resources should be used to maintain environmental quality. Every human activity that is in practice or sketched as a social requirement and for the sole purpose of the country's economic growth, either indirectly or directly, it influences the environment (Davis et al., 2009; Hernández et al., 2012). All over the world, the same effect is observed by the change in land resources, water resources, or any other resource, this change is mostly qualitative and sometimes quantitative too. Economic and social concerns may be different in under developing countries and developed countries (DCs) but in the environmental sector, the concerns are similar everywhere in the world, as every citizen requires safe and clean water for drinking and renewable energy (RE) which must be clean has to be supplied in sufficient quantity to each of them. Moreover, all activities which are agricultural and industrial rely on environmental resources which are common everywhere as deposits of minerals, forests, lands, rivers, soil, etc (Litfin, 1999; Corral-Verdugo et al., 2008).

There are some global environmental issues, while some are intra-regional and others are restricted to areas. For example, the economic condition and health of people may be affected in a negative way by the waste released in water, but this negative effect might be on the people of downstream countries. Similar to it is another example where agricultural land and health of people get affected by ozone depletion and this ozone depletion might be the result of transmission in the atmosphere by any other country which might be a hundred miles far. All countries are depending on natural resources which points out the reality that more coordination and collaboration between nations, on both regional and global scale, is required to meet environmental concerns as they are completely linked to poverty like economic issues (Jayadevappa and Chhatre, 2000).

Forced by poverty, people engage in activities that are destructive. For instance, they stay far away from utilizing services beneficial for their health, overuse the water, immoderate use of agricultural land and to get fuel, they utilize wood by chopping of trees. In a similar manner, social factors give rise to environmental issues such as increasing population results in more utilization of natural resources as it increases the fundamental needs of food, water, healthcare, electricity, sanitation, and housing.

2.6. PROBLEMS RELATED TO ENVIRONMENT AND INTERNATIONAL TRADE

Property rights and their importance under the light of international trade are studied here. Let's assume a model of two countries which are alike, having same priorities, using same technologies, utilizing similar endowments and giving similar inputs and generating same outputs, if these countries take part in free trade as a competitor in markets which are not regulatory, they will be different only in terms of proprietorship of environmental resources that they are utilizing as an input. Considering this case, common ownership problems can be the worst result of such trade which was initiated with a lack of rights in terms of property. For two alike countries, the efficiency of trade is not necessary; still, trade takes place only because they do not have the same property rights regimes. Under these circumstances, two propositions are accepted (Scott, 2004). According to the first one, the country who has defined its property rights vaguely or unclear, perceives that production is relying on the environment, they consider the environment as input and these vaguely defined rights, own their own make a ground for trade among alike countries. According to the second one, the country with improperly designed rights sets trade with a country that has properly designed rights, leads to miscalculation, and more use of resources, and the whole world economy gets affected by it. The allocation of resources does not become better by trade, though trade is helpful in equalizing the prices of goods traded. One country produces more resources than the required while another country consumes more resources than the required, so resources are underpriced. World Bank's document named Lawrence Summer's leaked and it gave rise to the debate about whether the countries which are developing have a relative edge in "dirty industries." It is truly not efficient, if they particularize in "dirty industries" and also if they are rigorous in environmental production (Subedi, 1999; Andonova et al., 2007). The relative advantages observed apparently are not genuinely relative advantages as they do not comprise comparative abundance in terms of resources, similarly, they do not have different preferences in productivity, and similarly, factor prices are not low. In accordance with this context, under developing countries are producing and exporting goods which are environmentally intensive to a greater extent and they are also at low social costs. Therefore, the under-DCs, by particularizing in "dirty industries" do not become wealthy nor do the world. Pareto optimality welfare concept also contradicts with this approach (Taskin and Zaim, 2001).

2.7. ENVIRONMENT AND ECONOMIC GROWTH, TRADE-OFF

Environmental pollution is caused by more usage of resources and this more usage is due to the wish of higher growth, hence between these two, the environment, and the economic growth, there exists a trade-off. Poor people of under-DCs solely rely on fuel on forests, for water on rivers, and for food on the soil. As they are left with no other option other than destroying natural habitat not by overusing them but by using them to survive as they desperately, need resources (Beckerman, 1992; Israel and Levinson, 2004).

By doing this, they risk their children as well as their own life. Environmental destruction expands by increasing the level of consumption as well as the level of income of the poor. To settle the more demand and consumption and at the same time ensuring that environmental degradation remains at the bottom level is a great challenge. Environmental degradation will be observed as these countries will utilize more resources in order to meet the wish of economic growth. The economical growing is a key element for providing more opportunities to poor countries; their developing model has to be environmentally sounder and should use less energy. On the other hand, industrial countries still cannot stop the growth and even cannot slow it down, as the global environment needs to be protected, so this option of stopping never exists. If they slow the growth, the poor nations would have to compromise as they depend on the rich nation's market (Suri and Chapman, 1998). In addition to it, to generate technologies that are safe for the environment and also for providing a margin of resources so that they can be transferred to poor people, the rich nations need to pursue their growth.

The model of growth of industrial states needs to be changed significantly. Quality needs to be exchanged with growth quantity. To establish stability between degradation and economic growth, less developed countries (LDCs) environmental destruction, and poverty, the cycle between these two needs to end. DCs protectionism is responsible for the decline in under-developing countries (UDC) earning capacity. A developing country not being able to access the goods market experiences the annual loss, which was almost double the amount, received as an aid in 2000 by developing countries, as estimated by the United Nations in 2001. The yearly loss may become $500 billion if the labor market and lack of capital access are added. The countries with industries are assembling countries that are being developed; this is done by providing them subsidy in their agricultural sections, according to an estimate which is $300 billion per year (Haq, 1999). Policies of industrial countries need to be changed so that the poverty circle of developing countries can be ended (Arrow et al., 1995; Stern et al., 1996).

2.8. EXTERNALITIES IMPACT ON ENVIRONMENT

From the above example, we can conclude that people try to balance their personal costs against personal benefits, just to boost up their personal welfare. Low priced coal being used to generate electricity imposes the reduced cost on others; this reduced cost is in the shape of water and air pollution. Such costs have no link to the loss or profit of the firm, so it is external to it. These costs are nothing less than real, as society is itself responsible, birth to them is given by society, not by the firm. Resources are not allocated by the firms, especially when external cost exists there. Not firm but the society is actually allowing the usage of valued resources such as water and air at no cost to by power plants (Boyle and Kiel, 2001; Söderholm and Sundqvist, 2003). Private costs and social costs difference can be demonstrated as a market arrangement's efficiency. Specific producers sustain the resources which are termed as private cost, while contrary to it; the overall cost of resources utilized only in the production section is termed as social cost (Lopez, 1997). The external cost comes in the act when both, the social and private costs are different. An external cost is actually equal to that difference between these two.

External costs = Social costs-private costs

If we consider external costs like pollution, goods produced by firms might be much more but as an effect, pollution in the environment will increase at the same rate. We can say when the allocation of resources is not efficiently allocated, there exists a social cost. The environment gets negatively affected by externalities (Van Woensel et al., 2001; Osterberg and Wallinga, 2004).

2.9. ENVIRONMENTAL MODELS

To estimate the extent to which environmental problems are present and to get able to mount a set of strategies to rule out programmers, for the betterment of billions of people, as developing countries face air pollution and water deficiency, empirical studies are being carried out, all over the world. These studies are funded by the European Union, Asian Development Bank, World Bank, and other Regional Institutes. Indoor combustion and as well as outdoor combustion produces particulate matter of the size 2.5 diameter, it is believed to cause more damage than other air pollutants (Hamby, 1994; Holgate *et al.*, 1999).

Researchers have pointed out that resource degradation and poverty in rural areas determine people's vulnerable frangibility of land (World Bank, 2003; Bennett et al., 2013). Fragile lands are usually steep, slopped or mostly forests are covering them. People living on fragile lands are mostly poor and are estimated to be 1.4 billion. Both the policymakers and researchers agree that these people do bear a great risk of neediness, better said to be impoverishment and resource degradation. People of Asia, especially the ones in East Asia and South Asia face this problem more. In both these places, deforestation leads to a loss of 1.2% forest, which is said to be highest in tropical regions. Studies have traversed the relation between the quality of the environment and economic development. Subsiding efforts, pollution damage, suppositions made for development about the welfare functions, subsiding cost, track for pollution and production of capital, all these are discussed in these theoretical papers which are written by Selden and Song (1995) and John and Pecchenino (1997).

Hettiqe (1997), Shafik (1994), Selden and Song (1995) carried out empirical studies and searched for systematic relations by reverting back to basics, to the air that passes all over the country, quality of water and income per capita specifications for various polynomials (Pedley, 1990; Jakeman et al., 2006).

2.9.1. Kuznets Environmental Curve Model

Simon Kuznets was a U.S. economist, suggested that when development proceeds, inequality in income usually rises, as the makeup of growth accumulates (Samimi et al., 2011). It has been claimed to pinpoint "Kuznets Environmental curve Model," in which overall all pollution whether from vehicles, houses or from industries increases until that much wealth through development is generated that proper measures for pollution control can be promoted (Kos' Yan et al., 2001; Awan, 2013; Gaur, 2018). This hypothesis proposed by Kuznets was based on per capita income. Nevertheless, controversy exists among many economists. This controversy arises about the point where the line turns, the country's per capita income, in case of inequality or environmental pollution initiates to decrease (Wesselung et al., 1996; Granell et al., 2010). It leads to ambiguity and as a result, it raises questions: Turnaround occurs at which income level? Is a similar trajectory followed by all sorts of pollutions? Whether the "Kuznets" curve is followed by pollution or not, if followed, then in the same manner? Or it falls first as income increases and rises later? Is structural change responsible for developed economies' reduction in pollution? (Figure 2.1).

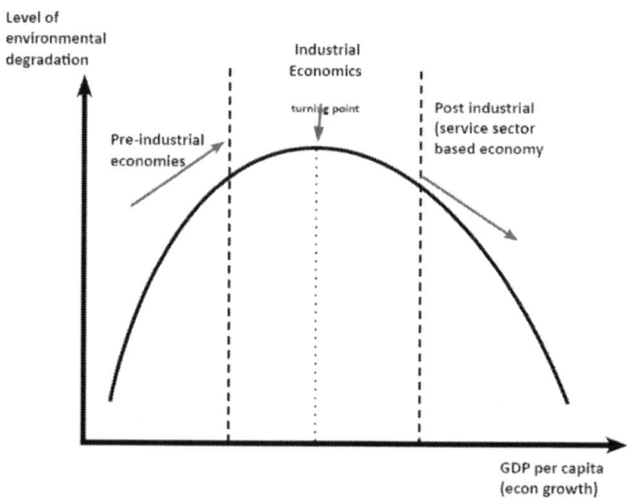

Figure 2.1: Environmental Kuznets curve.

Source: https://www.economicshelp.org/blog/14337/environment/environmental-kuznets-curve/.

Water pollution results are mixed but for air pollutants like sulfur dioxide and other suspended particulates, the Kuznets curve gives roughly consistent empirical results.

Kuznets Model is also confronted by emission studies (CO_2). These studies are considered hypothetical and according to the increase in vehicular emission are due to fact that more personal cars are purchased and utilized due to an increase in income, this is basically an effect of per capita income increase.

Empirical studies have further classified industrial pollution in the following three categories:

- National output;
- Industry's total share in national output; and
- Polluting sectors share.

Reducing the intensity of pollution in polluting sector and other combinations from discussed three factors which may result in changes related to income like the share in industrial output contributed by polluting sector and industry's share in the output of the nation, these combinations need to be processed in a way that they can decline pollution. According to an econometric result, which was carried out to measure economic development on water pollution by industries, it showed that between these two, the income and discussed three factors, there exists no U-shaped inverse relation. Joint product and income, both have different relations with discussed three factors. The outcome is not parabolic, instead, it is asymptotic. Unless the middle status of income is attained by the country, water emission by industries rises, and after that when a country grows richer, the line appears to be constant. These results show that output in reference to labor intensity and pollution continuously declines. This declining rate is similar to the rate at which income increases. While in the process of development, approximate consistency is observed in the ratios of sectoral pollution/labor. Nonetheless, the hypothesis suggested by Kuznets on water pollution by industries, this hypothesis reveals regularity in the environmental performance of a country.

2.9.2. Mehboobul Haq's Model

The idea is founded on a model that is constructed on SD. The focus of this novel idea is on primary concerns of people, investors must plan to include new technologies that are safe for the environment, and while making decisions, they should look for ways that can protect the resources, which can overcome the shortage of natural resources. There are five pillars of this model:

According to the first pillar, people must be placed at the center by the model. Protection of the environment is surely a vital thing, yet alone, it is not as human options, and protection of human life has to be the prime aim. Any step taken for the environment should be tested on the base that what it adds and how much benefit it introduces and also that how much welfare it adds to the world's population. We can say, human development must be the first pick (Ahmed et al., 2017).

According to the second pillar, the model has to be necessarily found on safe and sound technologies for the environment, especially in the energy sector. Particulate emission can be reduced to 99% simply by choosing energy technologies that are clean (unleaded fuel, technologies using clean coal and natural gas, etc.). Change in the pattern of the energy being consumed is necessary. For instance, energy consumed by the world can be reduced by 36%, if industrial countries follow the energy efficiency level of Japan. The energy utilized in kgs of oil equal per $100 of GNP, in Germany is 18, in the United States are 35, in Canada are 50, in Chine are 187, and in India are 132. Japan sets the lowest of 18. Introducing and adopting new patterns of technologies can speedily reduce energy consumption and in this way, they will serve as a curtain against emissions that are damaging the environment.

According to the third pillar, economic agents require incentives that are powerful enough to reflect the righteous utility in terms of decision-making for the environment. A system should be developed to control the pricing of resources.

According to the fourth pillar, the proposed model must fulfill two requirements, which are, it should be community-based, and it should be participatory. As it is believed that no one is useless, everyone must have got something precious to contribute, so all civil society sectors should be mobilized by the model. Natural resources must be controlled by communities living in that area. Grasslands, wildlife, fisheries, local wisdom, norms, and experiences of people must be drawn. Till that day, when local communities start to show interest and put in contribution to sustainability, no model can be practically implemented.

According to the fifth pillar, north, and south are two general divisions of the world. South comprises the poor countries whereas rich countries are mostly located in the north. 60% of food, 85% wood, 75% metal, and 70% energy are consumed by the north, though it has only one-fifth of the total population of the world. For replication of the same standards as used by the north, the south will have to consume 10 times more fossil fuels than being used in present and almost 200 times more wealth of minerals than being used now. An elementary change is required to distribute wealth and global consumption. The way north consumes material, its pattern cannot be duplicated everywhere in the world. So is the case with income, 85% of global income by north cannot be entitled forever. The planet can carry limited capacity of materials, this gives rise to questions associated with the lifestyles of rich people, also it questions simple and less energy-consuming lifestyles overall in the world. A development opportunity needs to be redistributed.

2.9.3. Three-Stage Development Model

As suggested by researchers, the development process is divided into three stages so that its impact on the environment can be measured. In populated areas, usually the urban ones, the decline in environmental quality and fast growth in industries is mostly witnessed at the starting stage of economic development. Intensively polluting emission from new industries results in increasing atmospheric pollutant concentration to a level that it becomes harmful (Phinney, 1993; Moini, 1995). This phenomenon of decomposing industrial emission is explained by the equation given below:

$$P = m(y)\, Q\, p(y)\, n(y)$$

where: P = total industrial pollution; m = manufacturing share in total output; Y = per capita income; Q = total output; P = manufacturing pollution intensity; and n = degree of pollution abatement: $o < n < 1$.

Hypothetically, three parameters involved in the equation are economic development functions: total output of shares generated by manufacturing (m), the degree to which pollution is abated by industry (n), and the manufacturing pollution intensity (p). The impact of economic development, in decomposition, relies on the magnetite and signals of parameters that are responsible for governing different relations among y, m, n, and p. Studies have proven the relation in economic development and industrialization to be an inverted-U relation between both, per capita income (y) and output of shares generated by manufacturing (m). Expanding industries rapidly than agriculture also increases economic growth, this happens in the initial phase

(Xiao, 2011). The observed growth in services serves as a dominating factor, this happens when the economy matures and as a consequence, m declines. Industries do significantly leave an impact on sectoral composition; this effect of industrial activity is on its intensity of average pollution or in terms of per unit of output. For instance, cement, and metal both are air pollutants which are intensively harmful, paper, and food production are not appropriate emitters (water pollutants) (Hettiqe, 1997; Attia et al., 2005). Cement, textile, metal, and sugar are primary industries. These industries transform mass raw materials into prime inputs. When in the developing phase, these primary industries probably can lose the share of output against cleaner industries. Electronics, assembling, and autos, etc industries are all cleaner industries (Bonnemaison and Marcoux, 1990; Fawcett et al., 2008).

2.10. IMMODERATE UTILIZATION OF RESOURCES

In recent six decades, great stress has been experienced by the environment and it is due to the tampering of humans at a massive level, this tempering is with the world's interconnected web of life, along with water, land, and air. Climatic change, species loss, and deforestation inflicting environmental damage are also coupled among previously discussed challenges. The quantitative decline is the result of immoderate and poorly planned utilization of natural resources. For example, catch per effort is a result of over-fishing and in aquatic systems, as a consequence change in the balance of ecological species (Al-Ibrahim, 1991; Kim, 2010). Insufficiency of clean water available for drinking in many cities is a result of over-exploitation of water. In coastal areas, the Stalinization and intrusion of seawater are caused by groundwater's overexploitation in an unscientific manner. Deforestation on a large extent has resulted in the loss of biodiversity. This loss includes economically important animal species and plant species. Soil erosion is a consequence of huge vegetation loss; it has also degraded the quality of the soil. Deforestation has also increased the sedimentation rate is stagnant. Environmentally important issues like global warming, change in climate, depletion of ozone, and eutrophication of aquatic systems, all are raised by industrialization (Shortle and Miranowski, 1987).

2.11. DEVELOPED COUNTRIES (DCS) CONTRIBUTION IN ENVIRONMENTAL POLLUTION

About 25% of resources are used by the United States, which has only a 2% population of the world. Similar is the case with European countries, extensive resources are exploited by a small population. More chemicals, more waste, and more hazards are generated by these countries as they use more resources. Only to avoid strict laws and for a safe environment, environmental hazards are exported by these countries to under developing countries in the following manner (Holt, 1995; Akbostanci et al., 2007).

2.11.1. Dirty Industries Shifted

The existence of pollution generating industries is prohibited by environmental laws. So illegally, DCs shift chemical hazards and these industries to under developing countries. World Bank Chief Economist, Lawrence Summer wrote a memo which was leaked by The Economist magazine, in 1992. This memo writes as, "Encouraging more shifting of dirty industries" like a ship breaking to LDCs like Pakistan, India, and Bangladesh where 90% of the world's ships are dismantled. It is estimated that in seawater, there are almost 45000 ships. Every year around 700 ships become obsolete. Hence, they need disposal. Till the 1970s, the breaking and disposal of the ship were done in the country but later due to more labor cost and a prominent increase in environmental regulations, such industries were shifted to under developing countries (Lazarova et al., 2001; Sulemana et al., 2017).

ILO suggests that asbestos, mercury, arsenic chromates, and lead are the materials that usually scrapping releases. The toxin export was banned under the 1992 Basel Convention. Hence, exporting these chemicals is surely a violation of this convention. These man-killing chemicals are still being exported to under developing countries as DCs are least bothered with these international laws (Jalil and Feridun, 2011; Marynowski et al., 2015). It is estimated that around 365 people die yearly during the process of shipbreaking.

2.11.2. Dumping of Chemical Wastes

US Company named Stroller Chemical Co (SSC) exported waste of nearly 1000 tons which was declared hazardous because of intensive cadmium and lead. Mixed in fertilizers, illegally, it was exported to Bangladesh. The government received no notification about it.

Moreover, "Flashier" is an American ship, reached the Bay of Bengal so that it can dump waste which was toxic and hazardous; luckily dumping was prevented as media jumped in a timely. In accordance with medical experts, the brain is damaged and children's growth is affected by chemicals such as cadmium and lead. Dumping these chemicals in water is a great threat to wildlife, as well as to the environment. It also imperils the world temperature control system. For both, the natural resources and human health, chemical waste is a great endanger.

2.11.3. Dumping of Electronic-Wastes

Among all manufacturing industries, the electronic industry is one of the most rapidly growing industries and for this reason; fast product obsolescence is also observed. E-waste is a term used for discarded electronic items. In the industrial world, E-waste is now considered as the most rapidly growing waste stream. Its increasing quantity is reaching catastrophic proportions; all industrial countries are facing a great struggle in fighting against this problem. Scrap wastes from computers are being dumped by many countries including the USA, UK, Kuwait, Saudi Arabia, U.A.E, Australia, and Japan. Serious contamination problems are faced when computer junk piles up as it emits toxic air as well as toxins in the water. Hence, serious recycling issues related to E-waste are there.

2.12. ENVIRONMENTAL HAZARDS AFFECTING HUMAN HEALTH

In the last 25 years, Hepatitis has grown to an alarming level; we have already talked about the data in the introduction section. According to Paul Russell, other than a disease, there is no such thing that is more international. In 1999, Haq states that Hepatitis like diseases travel continent to continent, from country to country, they do not follow any geographical restriction and they travel even without any sort of passport, this is because geographical and political boundaries are not followed by disease and health. Diseases are constantly a threat to other parts of the world. HBV is observed to be the second major cause of cancer after tobacco (24). Hepatitis B is a huge danger to health. According to reports of surveys, almost 30% population which comprises 350 million people who have symptoms of this infection, and out of the 1 to 2 million people face death each year. All these deaths are due to HBV related infection (Soine, 1987; Cabello and Godfrey, 2016).

Disease like hepatitis B is indigenous all over the world. In some European regions, mostly in under developing and tropical countries, this disease is spreading widely. Its widespread totally depends upon the country, it might be different in each country, and also it depends on the mixture of host factors and environmental behavior. Countries that have high living standards such as America, Europe, and Australia experience fewer diseases, but in countries with poor socioeconomic conditions, the disease rate is highest. The countries with higher disease rates include Africa, South America, and Southeast Asia. It is a contemptible and unfortunate condition that each year, out of 130 million, almost 30 million babies, all around the world do not access immunization (Rich et al., 1995; Adeola, 1994).

Since 1993, vaccines against hepatitis B are being recommended by the World Health Organization (WHO). Still, 1 million individuals are killed each year by it. There is a carrier rate of less than 1% in industrial countries. 90% of persistent carriers of HBV, from a total of 350 million people, live in under developing countries. Developing countries are on major threat of this disease. As indicated by a research report, Pakistan's 4.5% population which means almost 7 million individuals are affected by hepatitis B. Compared to AIDS, HBV is dozens of times more fatal, every tenth person in Pakistan is carrying germs of this disease. In accordance with the WHO Report, 2006, HIV has affected 39 million people by 2005, among them, half were women. Women (in Africa and the Caribbean) aged between 15 to 24 are likely to get infected almost six times more than men of the same age. Social factors like gender discrimination might be the key reason. Women need to lower the risk of infection. A better literacy rate may help women in adopting protection strategies. Around, 5 million new cases of HIV were reported in 2005, the reported cases bring the number of people affected by this virus to be approximately 40.3 million. The rising trends are highlighted in Figure 2.2. It was zero in 1980 and after a short time span of 25 years; 80 million people have been affected, which surely makes it a frightening concern (Brulle and Pellow, 2006).

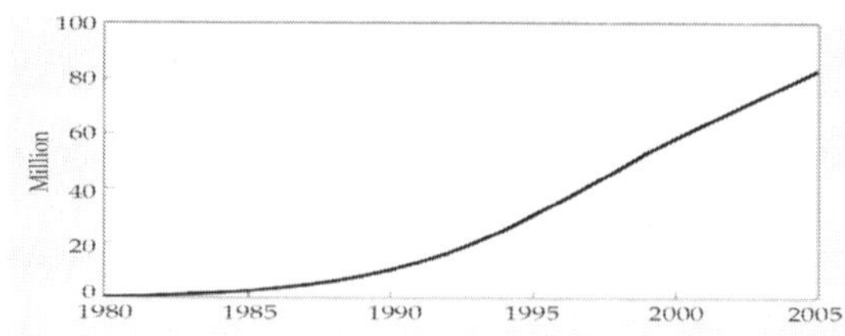

Figure 2.2: Estimates of global HIV infections (1980–2005).

Source: https://www.aessweb.com/pdf-files/741-761.pdf.

In 2005, due to AIDs, around 3 million individuals died. 0.6 million from a total of 3 million were children (less than 15 years). Since 1981, approximately 37.1 million individuals faced death due to AIDs. In last year, nearly 3 million people, (64% new infection) were affected in the region of Sub-Saharan Africa. The Caribbean is considered to be the second number in the list of AIDs affected region. Around 66% of young women lack knowledge about the transmission of HIV, as reported by the survey which was done in 24 countries of Sub-Saharan Africa. The high infection rate in women increases the infection risk in children too. During pregnancy or due to breast-feeding, children get infected through mother. In 2005, almost 2 million children (less than 14 years) are infected with HIV. We can conclude that poverty, inequality, and violence disproportionately affect women.

2.13. ENVIRONMENT AND MARKET INCENTIVES

On this occasion, the question that pops in is why firms and consumers tend to pollute water, land, and air? There are certain market forces. Usually, an assumption is made that these forces are responsible for influencing the behavior of firms, government agencies, and also of people consuming it. An elementary role is played by the incentives. Consumer habits can be changed by providing them a reduced price, the incentive can make up for the reduced price. Similarly, providing a high margin in profit in terms of incentive will surely encourage production, both in terms of services as well as goods. Market incentives can also be offered in a form that allocation of resources becomes efficient. Accordingly, it will not be surprising to know that these incentives play a vital role in the behavior of pollution (Smucker et al., 2005; Wang et al., 2008). We all know that thermal power plants are a significant cause of air pollution and water utilized in generating thermal power is polluted too. If one is the owner of the plant and he strives to make a maximum profit which is only possible by max production. One will be keeping marginal cost (MC) and revenue equal. Usually, electricity prices are fixed. Electricity production associated with the average total cost and MC is showed in the following Figure 2.3(a). At point A, equating marginal revenue (price) to MC, we can see maximization of output as a plant generates 1000 KW hours per day. Shaded rectangle between the average total cost curve (ATC) and price line illustrates total profits. Cheap availability of fuel such as high sulfur coal is the reason behind achieving a high-profit level. High-cost fuels like low sulfur coal and others, the one used in nuclear reactors are prime villains (Ackerman and Stewart, 1987).

Considering environmental aspects, plants are shifted from low-cost fuel to fuel that is high in cost, MC curve, and ATC both will shift upward (to MC2 and ATC2). Figure 2.3(b) shows it. The profit and output both will reduce. The output will be now decreased to 800 KWT-hours per day from 1000 KWT (Fershtman, 1985; Lacireno-Paquet et al., 2002). As output and profit have reduced, a firm who was experiencing maximum profit, will that firm would be satisfied with a higher cost, will they do it just to eliminate pollution emitted from plant to provide benefit to people living downstream (Dwyer, 1993; Russi et al., 2016). We really do not think that a firm generating maximum profit will even think about taking measures for environmental concerns. Not by aesthetic concerns, or for human welfare or by generosity but only on the base of cost and revenue comparison, the firm will consider the profit-maximizes behavior.

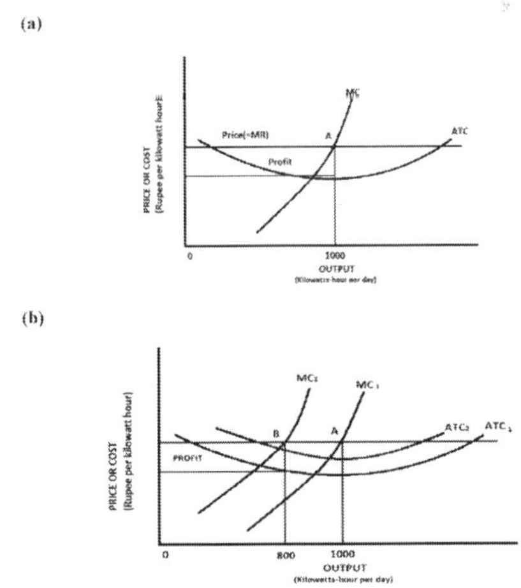

Figure 2.3: ATC and MC curves for high-cost fuels.

Source: https://www.aessweb.com/pdf-files/741-761.pdf.

REFERENCES

1. Ackerman, B. A., & Stewart, R. B., (1987). Reforming environmental law: The democratic case for market incentives. *Colum. J. Envtl. L.*, *13*, 171.
2. Adeola, F. O., (1994). Environmental hazards, health, and racial inequity in hazardous waste distribution. *Environment and Behavior*, *26*(1), 99–126.
3. Ahmed, W., Haq, M. U., Ali, S., Alvi, B. A., & Anis, M. I., (2017). Performance analysis of short distance wireless sensor networks using mote view. *Asian Journal of Engineering, Sciences and Technology*, *7*(2).
4. Akbostanci, E., Tunc, G. I., & Türüt-Aşik, S., (2007). Pollution haven hypothesis and the role of dirty industries in Turkey's exports. *Environment and Development Economics*, *12*(2), 297–322.
5. Al-Ibrahim, A. A., (1991). Excessive use of groundwater resources in Saudi Arabia: Impacts and policy options. *Ambio*, *1*, 34–37.
6. Altieri, M. A., Funes-Monzote, F. R., & Petersen, P., (2012). Agroecologically efficient agricultural systems for smallholder farmers: Contributions to food sovereignty. *Agronomy for Sustainable Development*, *32*(1), 1–13.
7. Andonova, L., Mansfield, E. D., & Milner, H. V., (2007). International trade and environmental policy in the post-communist world. *Comparative Political Studies*, *40*(7), 782–807.
8. Ang, B. W., Zhang, F. Q., & Choi, K. H., (1998). Factorizing changes in energy and environmental indicators through decomposition. *Energy*, *23*(6), 489–495.
9. Arrow, K., Bolin, B., Costanza, R., Dasgupta, P., Folke, C., Holling, C. S., & Pimentel, D., (1995). Economic growth, carrying capacity, and the environment. *Ecological Economics*, *15*(2), 91–95.
10. Attia, A. M., Honeycutt, Jr. E. D., & Leach, M. P., (2005). A three-stage model for assessing and improving sales force training and development. *Journal of Personal Selling and Sales Management*, *25*(3), 253–268.
11. Awan, A. G., (2013). Relationship between environment and sustainable economic development: A theoretical approach to environmental problems. *International Journal of Asian Social Science*, *3*(3), 741–761.

12. Azapagic, A., (2004). Developing a framework for sustainable development indicators for the mining and minerals industry. *Journal of Cleaner Production, 12*(6), 639–662.
13. Bansal, P., (2005). Evolving sustainably: A longitudinal study of corporate sustainable development. *Strategic Management Journal, 26*(3), 197–218.
14. Barrett, S., & Graddy, K., (2000). Freedom, growth, and the environment. *Environment and Development Economics, 5*, 433–456.
15. Beckerman, W., (1992). Economic growth and the environment: Whose growth? Whose environment? *World Development, 20*(4), 481–496.
16. Bennett, N. D., Croke, B. F., Guariso, G., Guillaume, J. H., Hamilton, S. H., Jakeman, A. J., & Pierce, S. A., (2013). Characterizing performance of environmental models. *Environmental Modeling and Software, 40*, 1–20.
17. Benotsch, E. G., Brailey, K., Vasterling, J. J., Uddo, M., Constans, J. I., & Sutker, P. B., (2000). War zone stress, personal and environmental resources, and PTSD symptoms in Gulf war veterans: A longitudinal perspective. *Journal of Abnormal Psychology, 109*(2), 205.
18. Bockstaller, C., & Girardin, P., (2003). How to validate environmental indicators. *Agricultural Systems, 76*(2), 639–653.
19. Bonnemaison, M., & Marcoux, E., (1990). Auriferous mineralization in some shear-zones: A three-stage model of metallogenesis. *Mineralium Deposita, 25*(2), 96–104.
20. Boyle, M., & Kiel, K., (2001). A survey of house price hedonic studies of the impact of environmental externalities. *Journal of Real Estate Literature, 9*(2), 117–144.
21. Bromet, E., & Moos, R. H., (1977). Environmental resources and the post-treatment functioning of alcoholic patients. *Journal of Health and Social Behavior*, 1, 326–338.
22. Brulle, R. J., & Pellow, D. N., (2006). Environmental justice: Human health and environmental inequalities. *Annu. Rev. Public Health, 27*, 103–124.
23. Cabello, F. C., & Godfrey, H. P., (2016). Even therapeutic antimicrobial use in animal husbandry may generate environmental hazards to human health. *Environmental Microbiology, 18*(2), 311–313.
24. Corral-Verdugo, V., Carrus, G., Bonnes, M., Moser, G., & Sinha, J. B., (2008). Environmental beliefs and endorsement of sustainable

development principles in water conservation: Toward a new human interdependence paradigm scale. *Environment and Behavior*, *40*(5), 703–725.

25. Davis, J. L., Green, J. D., & Reed, A., (2009). Interdependence with the environment: Commitment, interconnectedness, and environmental behavior. *Journal of Environmental Psychology*, *29*(2), 173–180.
26. Dwyer, J. P., (1993). The use of market incentives in controlling air pollution: California's marketable permits program. *Ecology Law Quarterly*, *20*(1), 103–117.
27. Fawcett, S. E., Magnan, G. M., & McCarter, M. W., (2008). A three-stage implementation model for supply chain collaboration. *Journal of Business Logistics*, *29*(1), 93–112.
28. Fershtman, C., (1985). Managerial incentives as a strategic variable in duopolistic environment. *International Journal of Industrial Organization*, *3*(2), 245–253.
29. Gaur, A., (2018). Environment and sustainable economic development: Approach to environmental problems. *Journal of Industrial Relationship, Corporate Governance and Management Explorer (e ISSN 2456–9461)*, *2*(2), 23–26.
30. Giddings, B., Hopwood, B., & O'brien, G., (2002). Environment, economy, and society: fitting them together into sustainable development. *Sustainable Development*, *10*(4), 187–196.
31. Granell, C., Díaz, L., & Gould, M., (2010). Service-oriented applications for environmental models: Reusable geospatial services. *Environmental Modeling and Software*, *25*(2), 182–198.
32. Gregory, R., Lichtenstein, S., & Slovic, P., (1993). Valuing environmental resources: A constructive approach. *Journal of Risk and Uncertainty*, *7*(2), 177–197.
33. Hamby, D. M., (1994). A review of techniques for parameter sensitivity analysis of environmental models. *Environmental Monitoring and Assessment*, *32*(2), 135–154.
34. Haq, M., (1999). *Reflections on Human Development* (Vol. 1, pp. 1–20). Islamabad, Pakistan: Oxford University Press.
35. Harvey, L. D., (2013). Recent advances in sustainable buildings: Review of the energy and cost performance of the state-of-the-art best practices from around the world. *Annual Review of Environment and Resources*, *38*, 281–309.

36. Hernández, B., Suárez, E., Corral-Verdugo, V., & Hess, S., (2012). The relationship between social and environmental interdependence as an explanation of proenvironmental behavior. *Human Ecology Review*, 1, 1–9.
37. Hettiqe, H., (1997). *Pollution in Economic Development* (Vol. 1, pp. 1–20). Development Research Group, World Bank.
38. Holgate, S., Samet, J., Koren, H., & Maynard, R., (1999). *Air Pollution and Health* (Vol. 1, pp. 1–30). San Diego, California: Academic Press.
39. Holt, P. G., (1995). Review Environmental factors and primary T-cell sensitization to inhalant allergens in infancy: Reappraisal of the role of infections and air pollution. *Pediatric Allergy and Immunology*, *6*(1), 1–10.
40. Holt, R. P., (2005). Post-Keynesian economics and sustainable development. *International Journal of Environment, Workplace and Employment*, *1*(2), 174–186.
41. Hopwood, B., Mellor, M., & O'Brien, G., (2005). Sustainable development: Mapping different approaches. *Sustainable Development*, *13*(1), 38–52.
42. Israel, D., & Levinson, A., (2004). Willingness to pay for environmental quality: Testable empirical implications of the growth and environment literature. *Contributions in Economic Analysis and Policy*, *3*(1), 10–22.
43. Jakeman, A. J., Letcher, R. A., & Norton, J. P., (2006). Ten iterative steps in development and evaluation of environmental models. *Environmental Modeling and Software*, *21*(5), 602–614.
44. Jalil, A., & Feridun, M., (2011). The impact of growth, energy, and financial development on the environment in China: A cointegration analysis. *Energy Economics*, *33*(2), 284–291.
45. Jayadevappa, R., & Chhatre, S., (2000). International trade and environmental quality: A survey. *Ecological Economics*, *32*(2), 175–194.
46. John, A. A., & Pecchenino, R. A., (1997). International and intergenerational environmental externalities. *Scandinavian Journal of Economics, 99*(3), 1–30.
47. Kim, S. H., (2010). Morbid obesity and excessive hospital resource consumption for unilateral primary hip and knee arthroplasty. *The Journal of Arthroplasty*, *25*(8), 1258–1266.

48. Kos' Yan, R., Kunz, H., Pykhov, N., Kuznetsov, I., Podymov, I., & Vorobyev, P., (2001). Concentration and transport of suspended sediments in irregular swell. *Die Küste, 64*, 161–200.
49. Lacireno-Paquet, N., Holyoke, T. T., Moser, M., & Henig, J. R., (2002). Creaming versus cropping: Charter school enrollment practices in response to market incentives. *Educational Evaluation and Policy Analysis, 24*(2), 145–158.
50. Laffont, & Tirole, J., (1994). *Pollution Permits and Compliance Strategies* (Vol. 1, pp. 1–20). Working paper 95–9, Massachusetts Institute of Technology (MIT), Development of Economics.
51. Lamidi, R. O., Jiang, L., Pathare, P. B., Wang, Y., & Roskilly, A. P., (2019). Recent advances in sustainable drying of agricultural produce: A review. *Applied Energy, 233*, 367–385.
52. Lawn, P. A., (2001). Goods and services and the dematerialization fallacy: implications for sustainable development indicators and policy. *International Journal of Services Technology and Management, 2*(3/4), 363–376.
53. Lazarova, V., Levine, B., Sack, J., Cirelli, G., Jeffrey, P., Muntau, H., & Brissaud, F., (2001). Role of water reuse for enhancing integrated water management in Europe and Mediterranean countries. *Water Science and Technology, 43*(10), 25–33.
54. Litfin, K. T., (1999). Constructing environmental security and ecological interdependence. *Global Governance: A Review of Multilateralism and International Organizations, 5*(3), 359–377.
55. Lopez, R., (1997). Environmental externalities in traditional agriculture and the impact of trade liberalization: The case of Ghana. *Journal of Development Economics, 53*(1), 17–39.
56. Ma, Y., Chen, L., Zhao, X., Zheng, H., & Lü, Y., (2009). What motivates farmers to participate in sustainable agriculture? Evidence and policy implications. *International Journal of Sustainable Development and World Ecology, 16*(6), 374–380.
57. Mäler, K. G., (1991). National accounts and environmental resources. *Environmental and Resource Economics, 1*(1), 1–15.
58. Martin, B., McGutre, F., & Allen, L., (1998). Retirees' attitudes toward tourism: Implications for sustainable development. *Tourism Analysis, 3*(1), 43–51.

59. Martinelli, L. A., Naylor, R., Vitousek, P. M., & Moutinho, P., (2010). Agriculture in Brazil: Impacts, costs, and opportunities for a sustainable future. *Current Opinion in Environmental Sustainability*, *2*(5/6), 431–438.
60. Marynowski, M., Likońska, A., Zatorski, H., & Fichna, J., (2015). Role of environmental pollution in irritable bowel syndrome. *World Journal of Gastroenterology: WJG*, *21*(40), 11371.
61. Moini, A. H., (1995). An inquiry into successful exporting: An empirical investigation using a three-stage model. *Journal of Small Business Management*, *33*(3), 9.
62. Muzari, W., Gatsi, W., & Muvhunzi, S., (2012). The impacts of technology adoption on smallholder agricultural productivity in sub-Saharan Africa: A review. *Journal of Sustainable Development*, *5*(8), 69.
63. Nilsson, M., Griggs, D., & Visbeck, M., (2016). Policy: Map the interactions between sustainable development goals. *Nature*, *534*(7607), 320–322.
64. Olsthoorn, X., Tyteca, D., Wehrmeyer, W., & Wagner, M., (2001). Environmental indicators for business: A review of the literature and standardization methods. *Journal of Cleaner Production*, *9*(5), 453–463.
65. Omer, A. M., (2008). Energy, environment, and sustainable development. *Renewable and Sustainable Energy Reviews*, *12*(9), 2265–2300.
66. Osterberg, D., & Wallinga, D., (2004). Addressing externalities from swine production to reduce public health and environmental impacts. *American Journal of Public Health*, *94*(10), 1703–1708.
67. Pedley, H. M., (1990). Classification and environmental models of cool freshwater tufas. *Sedimentary Geology*, *68*(1/2), 143–154.
68. Peng, Q., Liu, Y., Gu, P., & Fan, Z., (2013). Development of an open-architecture electric vehicle using adaptable design. In: *Advances in Sustainable and Competitive Manufacturing Systems* (Vol. 1, pp. 79–90). Springer, Heidelberg.
69. Phinney, J. S., (1993). A three-stage model of ethnic identity development in adolescence. *Ethnic Identity: Formation and Transmission among Hispanics and Other Minorities*, *61*, 79.

70. Piorr, H. P., (2003). Environmental policy, agri-environmental indicators, and landscape indicators. *Agriculture, Ecosystems and Environment*, *98*(1–3), 17–33.
71. Prakash, C. S., (2000). *Scientists Release Declaration Defending Biotechnology: Genetically Modified Foods*. PR Newswire, Institute for Agriculture and Trade Policy (Vol.1, pp. 1-2).
72. Pretty, J. N., (1995). Participatory learning for sustainable agriculture. *World Development*, *23*(8), 1247–1263.
73. Rich, R. C., Edelstein, M., Hallman, W. K., & Wandersman, A. H., (1995). Citizen participation and empowerment: The case of local environmental hazards. *American Journal of Community Psychology*, *23*(5), 657–676.
74. Russi, D., Margue, H., Oppermann, R., & Keenleyside, C., (2016). *Result-Based Agri-Environment Measures: Market-Based Instruments, Incentives, or Rewards?* (Vol. 54, pp. 69–77) The case of Baden-Württemberg. Land Use Policy.
75. Samimi, A. J., Kashefi, A., Salatin, P., & Lashkarizadeh, M., (2011). Environmental performance and HDI: Evidence from countries around the world. *Middle-East Journal of Scientific Research*, *10*(3), 294–301.
76. Schaper, M., (2002). The challenge of environmental responsibility and sustainable development: Implications for SME and entrepreneurship academics. *Radical Changes in the World: Will SMEs Soar or Crash*, *1*, 541–553.
77. Schindler, J., Graef, F., & König, H. J., (2015). Methods to assess farming sustainability in developing countries: A review. *Agronomy for Sustainable Development*, *35*(3), 1043–1057.
78. Scholz, M., Hedmark, Å., & Hartley, W., (2012). Recent advances in sustainable multifunctional land and urban management in Europe: A review. *Journal of Environmental Planning and Management*, *55*(7), 833–854.
79. Scott, J., (2004). International trade and environmental governance: Relating rules (and standards) in the EU and the WTO. *European Journal of International Law*, *15*(2), 307–354.
80. Selden, T. M., & Song, D., (1995). Neoclassical growth, the j curve for abatement, and the inverted u curve for pollution. *Journal of Environmental Economics and Management, 29*(2).

81. Shafik, N., (1994). *Economic Development and Environmental Quality: An Econometrical Analysis* (Vol. 1, pp. 1–30). Special issues on Environmental Economics, Oxford University Press.
82. Shortle, J. S., & Miranowski, J. A., (1987). Intertemporal soil resource use: Is it socially excessive? *Journal of Environmental Economics and Management*, *14*(2), 99–111.
83. Smucker, G. R., Fleurantin, G., McGahuey, M., & Swartley, B., (2005). Agriculture in a fragile environment: Market incentives for natural resource management in Haiti. *Order*, (521-O), 00–05.
84. Söderholm, P., & Sundqvist, T., (2003). Pricing environmental externalities in the power sector: Ethical limits and implications for social choice. *Ecological Economics*, *46*(3), 333–350.
85. Soine, L., (1987). Expanding the environment in social work: The case for including environmental hazards content. *Journal of Social Work Education*, *23*(2), 40–46.
86. Stave, J. W., & Durandetta, D., (2000). GM crop testing grows amid controversy. *Today's Chemist at Work* (Vol. 1, 9, Nos. 6, pp. 1–9, 32, 33, 37). http://pubs.acs.org/hotartcl/tcaw/00/jun/stave.html (accessed on 1 May 2020).
87. Stern, D. I., Common, M. S., & Barbier, E. B., (1996). Economic growth and environmental degradation: The environmental Kuznets curve and sustainable development. *World Development*, *24*(7), 1151–1160.
88. Stevens, Jr. D. L., & Olsen, A. R., (2003). Variance estimation for spatially balanced samples of environmental resources. *Environmetrics*, *14*(6), 593 610.
89. Subedi, S. P., (1999). Balancing international trade with environmental protection: international legal aspects of eco-labels. *Brook. J. Int'l L.*, *25*, 373.
90. Sulemana, I., James, H. S., & Rikoon, J. S., (2017). Environmental Kuznets Curves for air pollution in African and developed countries: Exploring turning point incomes and the role of democracy. *Journal of Environmental Economics and Policy*, *6*(2), 134–152.
91. Suri, V., & Chapman, D., (1998). Economic growth, trade, and energy: Implications for the environmental Kuznets curve. *Ecological Economics*, *25*(2), 195–208.

92. Taskin, F., & Zaim, O., (2001). The role of international trade on environmental efficiency: A DEA approach. *Economic Modeling, 18*(1), 1–17.
93. Tennant, D. L., (1976). In stream flow regimens for fish, wildlife, recreation and related environmental resources. *Fisheries, 1*(4), 6–10.
94. Timur, S., & Getz, D., (2009). Sustainable tourism development: How do destination stakeholders perceive sustainable urban tourism? *Sustainable Development, 17*(4), 220–232.
95. Tripathi, V., Edrisi, S. A., Chen, B., Gupta, V. K., Vilu, R., Gathergood, N., & Abhilash, P. C., (2017). Biotechnological advances for restoring degraded land for sustainable development. *Trends in Biotechnology, 35*(9), 847–859.
96. Van Woensel, T., Creten, R., & Vandaele, N., (2001). Managing the environmental externalities of traffic logistics: The issue of emissions. *Production and Operations Management, 10*(2), 207–223.
97. Wang, Q., Wong, T. J., & Xia, L., (2008). State ownership, the institutional environment, and auditor choice: Evidence from China. *Journal of Accounting and Economics, 46*(1), 112–134.
98. Wesselung, C. G., Karssenberg, D. J., Burrough, P. A., & van Deursen, W. P., (1996). Integrating dynamic environmental models in GIS: The development of a dynamic modeling language. *Transactions in GIS, 1*(1), 40–48.
99. World Bank, (2003). *World Development Report 2003: Sustainable Development in a Dynamic World: Transforming Institutions, Growth, and Quality of Life* (Vol. 1, pp. 1–20). New York: Oxford University Press, U.S.A.
100. Xiao, L., (2011). Financing high-tech SMEs in China: A three-stage model of business development. *Entrepreneurship and Regional Development, 23*(3/4), 217–234.

CHAPTER 3

Energy and Sustainable Development

CONTENTS

3.1. Introduction 70
3.2. Energy and Social Issues 76
3.3. Energy, The Environment, and Health 80
3.4. Energy Security 85
3.5. Energy Resources 86
3.6. Energy End-Use Efficiency 88
3.7. Renewable Energy (RE) Technologies 91
3.8. Advanced Energy Technologies 92
References 100

3.1. INTRODUCTION

Energy end-use technologies and an energy supply sector combine to make an energy system. Delivery of the advantage offered by the energy is the main aim of the energy system. These advantages are described by the term energy services, in households; it includes cooked food, transportation, illumination, refrigeration, and comfortable indoor temperatures. Industrial activity and virtually every commercial energy also require energy services. For example, for many industrial procedures, cooling, and heating are required, for agriculture is required for motive power and electronics and telecommunications electricity is required (Adelman, 1972, 1995; ADB-African Development Bank, 1999).

Extraction or collection of primary energy is the beginning of the energy chain that offers these services that may be transformed into energy transporters in one or numerous steps such as diesel or electricity which are appropriate for the final uses (Ballin, 1946). Final energy is transformed into useful energy by light bulbs, machinery, stoves, and vehicles which are the end-use apparatus, which gives the required benefits, i.e., the energy services. An instance of the energy chain, which ends with the steel produced and beginning with the collection of coal from mine (primary energy source) as an energy service—is shown in Figure 3.1.

Various technologies, primary energy, infrastructure (capital), materials, and labor (know-how) combine to make the energy services. There are price tags for each of these inputs and are partially compatible with one another. The economic problems are utility resultant from the services or are the economic value, from the consumer's viewpoint. Upstream activities are often not known to the consumers which are needed to generate the services (Austin, 1998; ASTAE: Asia Alternative Energy Program, 1998).

In 1995, in the USA the primary energy consumption per capita was 330 gigajoules, this amount is more than eight times as much as consumed by the typical Sub-Saharan African (they used 40 gigajoules per year when traditional and commercial energy are counted). In developed countries (DCs), many people use much less. Commercial and non-commercial energy usage is shown in Figure 3.2 in several areas (Jefferson, 1983; Ito, 2000).

Numerous types of commercial energy are used by the affluent minority in most of the under-DCs in the same manner as most of the people in DCs. Inefficient use of technologies such as open fires or unaired stoves is the main source of non-commercial energy used by several people in under-DCs. Generally, in energy statistics, traditional energy sources are not considered.

Since the data collection of commercially distributed energy resources is readily available, thus analysis based on per capita consumption is common. However, the world's energy condition is now accurately reflected by the resulting analysis, which is the reason for the involvement of the estimates of non-commercial energy in Figure 3.3 and Table 3.1. Globally, non-commercial energy is very important, though less well recognized and is utilized far more extensively than commercial energy many developing countries in rural regions, mainly the least industrialized countries (Barland, 1999; Baranzini et al., 2000).

The energy that comes from the sun to the earth is thousands of times greater than the rate of global commercial energy intake. 80% of the total energy consumption is from primary energy sources such as fossil fuels (natural gas, oils, and coal) (Table 3.1). Renewables and hydropower deliver about 2% each and nuclear power delivers slightly more than 6%. About 10% of the total energy mix is attained from the traditional (mostly non-commercial) energy sources worldwide. But in industrialized countries, the distribution is not even: only 2% of energy consumption is accounted for from the non-commercial energy, but in developing countries, it accounts for 30%. Traditional biomass accounts for 90% in some low-income developing countries, of total energy consumption (IEA, 1997a, 1998, 1998b, c).

If the use of primary energy continues to grow about 2% a year globally, it will result in doubling the consumption of energy by 2035 compared to 1998 and will triple by 2055. In the past 30 years, commercial energy use has increased at a rate three and a half times in developing countries than that of OECD countries, by increasing personal incomes, the result of life-style variations made possible, along with the shift from traditional to commercial energy and h higher population growth rates. However, based on per capita, the rise in total primary energy use has not caused outstandingly in more reasonable availability to energy services between developing and industrialized countries. An excess amount of energy will be required to distribute prospects to the billions of people in developing countries who do not have access to satisfactory energy facilities and to fuel global economic development (Hughes, 1983; Hwang, 1995).

However, efficiencies by which the energy is produced, distributed, and used is responsible for the quantity of additional energy needed to deliver the energy services required in the future. In new energy supply systems, financial investments can be reduced by energy efficiency developments, as they have over the past 200 years. Across regions, between energy use and economic activity, the degree of interdependence is neither uniform nor static. The country's stage of development is often responsible for the energy intensity (the ratio of energy demand to GDP), which appreciate abundant energy facilities, development in energy requirement is less strongly connected to economic production than it was in the past (Figure 3.3).

The tendency towards a decrease in energy intensity as economic growth continues can be separated over a long-past era, as displayed in Figure 3.4, developing country examples of India and China are included. A thorough, long-term examination of energy intensity for several countries discloses a common pattern of energy use determined by the following features:

- The shift to commercial from non-commercial forms of energy, motorization, and industrialization originally raise the commercial energy-GDP ratio (This ratio enhanced in transition in economies in the 1990s, mainly because of slower economic development).
- As incomes increase and industrialization proceeds, saturation effects, and also the growth of the service sector (which is less energy-intensive), reduction in the ratio of commercial energy to GDP after it ranges a peak. Many countries have passed the maximum energy intensity but not by low-income developing countries.
- In the growing energy demand arising from growing production and incomes and increasing population, energy efficiency improvements can be the main limiting factor, as a result of world-wide technology transfer and diffusion.
- The more effective use of materials in well-designed, better-quality, miniaturized products, and the saturation of bulk markets for basic materials, the recycling of energy-intensive materials, industrialized countries contribute to added reductions in energy intensity.
- In developing countries, technological vaulting to the use of highly effective processes, transportation systems, machinery, appliances, and vehicles offers a substantial potential for energy efficiency developments.

In developing and industrialized countries, these drivers are causing a common pattern of energy use per unit of GDP.

Energy costs impact consumer behavior and choices and can influence economic improvement and development. High energy costs can prompt expanding import bills, with antagonistic outcomes for the business, work, and social welfare. High energy costs can likewise animate investigation and advancement of extra assets, make a pull for development, and give motivating forces to effectiveness improvements (Hippel, 1988; Houghton et al., 1990).

Although a few effects of energy costs are genuinely consistent, others are progressively transient. For instance, unique outright value levels have had little impact on economic development in OECD European nations or Japan comparative with the much lower energy costs in the United States and some developing nations. What influenced economic development in all energy bringing in nations was the value price hike of the 1970s.

Table 3.1: World Primary Energy Consumption, 1998

Source	Primary Energy (eJ)	Primary Energy (109 T Oil Eq.)	% of Total	Static Reserve-Production Ratio (Years)[a]	Static Resource Base-Production Ratio (Years)[b]	Dynamic Resource Base-Production Ratio (Years)[c]
Fossil fuels	**320**	**7.63**	**79.6**			
Oil	142	3.39	35.3	45	~ 200	95
Coal	93	2.22	23.1	452	~ 1,500	1,000
Natural gas	85	2.02	21.1	69	~ 400	230
Renewables	**56**	**1.33**	**13.9**			
Large hydro	9	0.21	2.2	Renewable		
'New' renewable[d]	9	0.21	2.2	Renewable		
Traditional biomass	38	0.91	9.5	Renewable		
Nuclear	**26**	**0.62**	**6.5**			
Nuclear[e]	26	0.62	6.5	50f	>> 300f	
Total	**402**	**9.58**	**100.0**			

[a] Based on constant static and fabrication reserves.

[b] Includes both unconventional and conventional resources reserves.

[c] Data denote the energy usage of a business-as-usual situation—that is, production is a function of demand and is dynamic. Thus under different scenarios, these ratios are subject to change.

[d] Includes modern biomass, wind energy, geothermal energy, solar energy, marine energy, and small hydropower. Modern biomass accounts for about 2 exajoules and 7 exajoules come from all other renewables.

[e] Converted from electricity formed to fuels consumed assuming a 33% thermal competence of power plants.
[f] Based on once-through uranium fuel cycles without thorium and low-concentration uranium from seawater. The uranium resource base is efficiently 60 times larger if fast breeder reactors are used.

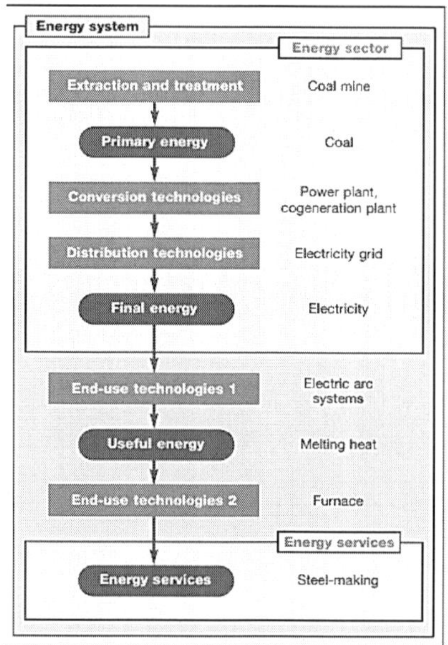

Figure 3.1: An example of the energy chain from extraction to services.

Source: https://www.undp.org/content/dam/aplaws/publication/en/publications/environment-energy/www-ee-library/sustainable-energy/world-energy-assessment-energy-and-the-challenge-of-sustainability/World%20Energy%20Assessment-2000.pdf.

The capital venture is essential for energy improvement. Energy framework improvement and basic change are the consequences of interest in plants, gear, and energy system infrastructure. Troubles in pulling in the capital for energy investment may obstruct economic advancement, particularly at all developed nations. Rare public funds, particularly in developing nations, are required for some ventures—going from provincial improvement, training, and health care to energy supplies. Since energy supply, more than some other option is regularly observed as more promptly capable of producing early revenues, energy investments are progressively seen as a private sector affair. However, private assets are not streaming into many developing nations for several reasons, particularly dangers to investors (Bowers, 1982; Bowers, 1982).

Foreign direct venture approached $400 billion out of 1997—up from $50 billion out of 1984—and speaks to an expanding portion of international investment flows. The foreign direct venture is commonly industrially inspired, and investors hope to recuperate the underlying capital as well as depend on competitive returns. These results cannot be ensured in developing nations with conceivably delicate governments or without free markets. Next to no outside, direct investment arrives at the least developed nations (Heede et al., 1985).

In contrast to foreign direct investment, official development aid has stayed level comparative with net world items. In 1997, it totaled $56 billion, or 0.25% of the GDP of OECD nations—which have concurred on a fundamental level to an objective of 0.7% of GDP.8 Against this scenery, financing is insufficient for energy extends in developing nations. Until the economic dangers to foreign investors can be overseen (for instance, through clear and stable standards for energy and money related markets, consistent income age through bill collection, and profit transfers), most developing nations may need to keep on financing their energy improvement from residential investment funds.

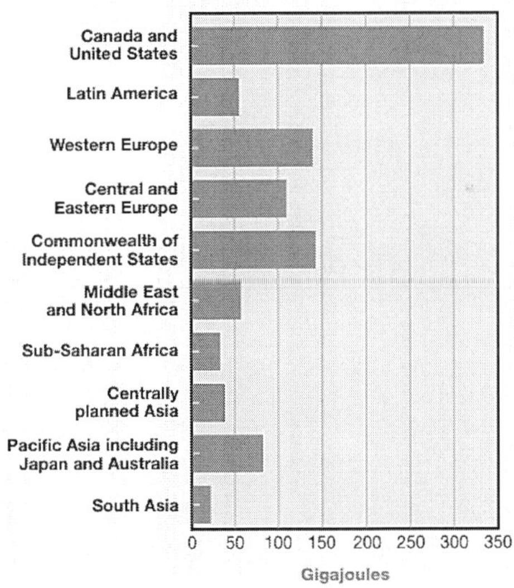

Figure 3.2: Primary per capita energy consumption (commercial and non-commercial) by region, 1995.

Source: https://www.undp.org/content/dam/aplaws/publication/en/publications/environment-energy/www-ee-library/sustainable-energy/world-energy-

assessment-energy-and-the-challenge-of-sustainability/World%20Energy%20 Assessment-2000.pdf.

3.2. ENERGY AND SOCIAL ISSUES

Energy use is firmly connected to a scope of social issues, including population growth, urbanization, poverty alleviation, and an absence of opportunities for ladies. Even though these issues influence energy demand, the relationship is two-way: the quality and amount of energy services, and how they are accomplished, affect social issues also (Hannah, 1982).

Neediness is the overriding social issue for developing nations. Some 1.3 billion individuals in the developing world live on under $1 per day. Salary estimation alone, nonetheless, does not completely catch the wretchedness and the absence of decision that neediness represents. The energy utilization examples of poor people—particularly their dependence on conventional fuels in rustic zones—will, in general, keep them ruined (Brenton, 1994; Brack, 1998).

Around the world, 2 billion individuals are without access to power and an equivalent number keeps on utilizing conventional solid fuels for cooking. As appeared in the following area, cooking with inadequately vented stoves has huge well-being impacts. Furthermore, a huge number of individuals—predominantly ladies and children—go through a few hours per day in the drudgery of get-together kindling and conveying water, regularly from considerable distances, for household needs. On account of these demands on their time and energy, women, and youngsters frequently pass up open doors for training and other productive exercises.

Absence of power, as a rule, implies deficient illumination and barely any labor-saving apparatuses, just as constrained media communications and opportunities for business ventures. More prominent access to power and present-day powers and stoves for cooking can empower individuals to appreciate both self-reinforcing and short-term, long-term advances in their satisfaction. Table 3.2 outlines a portion of the particular upgrades that may result (Hannah, 1979; Grubb et al., 1999).

Constrained salaries may drive family units to utilize conventional fuels and inefficient innovations. Figure 3.5 shows the normal primary energy interest for different powers as a component of income levels in Brazil. For low-salary family units, firewood is the predominant fuel. At higher wages, wood is supplanted by commercial electricity and fuels, which offer a lot more noteworthy comfort, energy effectiveness, and tidiness. Since affordable, convenient energy can add to a family's efficiency and income-creating potential, its accessibility can turn into a switch for breaking out of relative advantages and expenses of richness towards a lower number of wanted births in a family. A speeding up of the demographic transition to low fruitfulness and low mortality (as has happened in industrialized nations) relies upon significant formative undertakings, including educating women, improving the local environment, and ameliorating the extreme poverty that may make child labor a need. Every one of these undertakings has connections to the accessibility of ease energy services (Grübler, 1996).

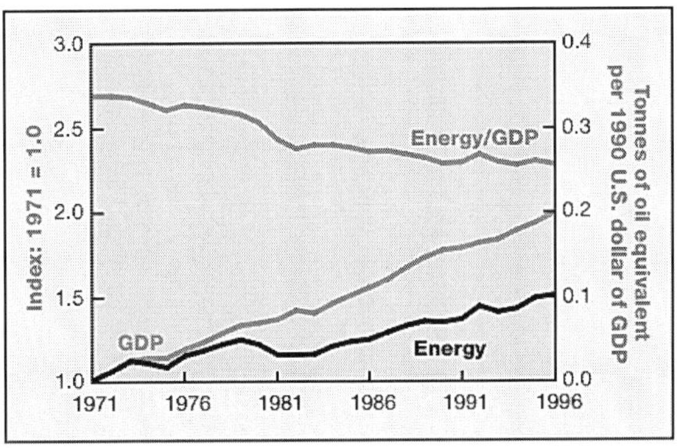

Figure 3.3: GDP and primary energy consumption in OECD countries, 1971–1996.

Source: https://www.undp.org/content/dam/aplaws/publication/en/publications/environment-energy/www-ee-library/sustainable-energy/world-energy-assessment-energy-and-the-challenge-of-sustainability/World%20Energy%20Assessment-2000.pdf.

The developing interest of people in urban focuses is another key segment issue connected to energy. Even though the general pattern towards urbanization has numerous segments and might be unavoidable, giving more choices to rustic occupants through energy intercessions, might slow relocation and diminish pressure on quickly developing urban areas. Even though the negative externalities related to energy use in urban regions can be serious, different procedures can alleviate their advanced energy protection. Thinking about vitality in land-use arranging, and in planning physical infrastructure, development norms, and transportation systems, can decrease a portion of the development in energy demand that goes with fast urbanization (Goldemberg et al., 1998).

Transportation systems might be particularly significant right now, the fast development in the number of motor vehicles around the world. Since around 1970, the worldwide fleet has been expanding by 16 million vehicles per year, and more than 1 billion autos will probably be out and about by 2020. The greater part of these vehicles will be driven in the urban areas of the developing world, where they will make more clog, disturb urban contamination, and undermine human wellbeing—even with hopeful projections about effectiveness improvements and substitute fuels (Goldemberg and Reid, 1998, 1999).

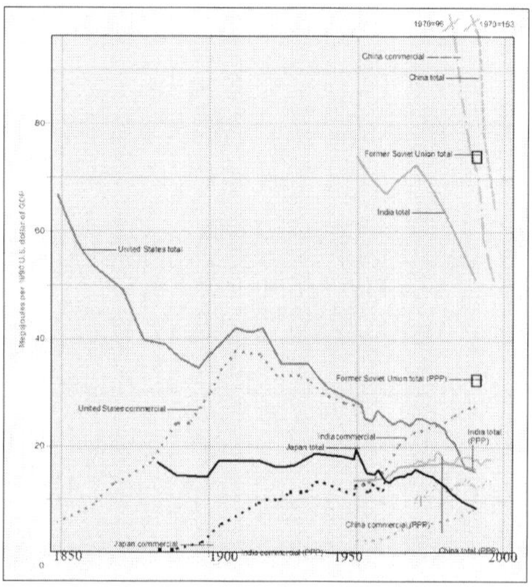

Figure 3.4: Two energy intensity paths are shown for Japan and the United States, one based on total energy consumption from all sources and the other only on commercial energy.

Note: The paths converge where traditional sources have been replaced by commercial energy. Because of distortions from market fluctuations, energy intensity paths for China and India are calculated in two ways: using total and commercial energy divided by GDP measured at market exchange rates (as with Japan and the United States), and divided by GDP measured at purchasing power parities (PPP). Energy intensities for the former Soviet Union derived using both market exchange rates and PPP, are data points only (Naki´cenovi´c, Grübler, and McDonald, 1998).

Source: https://www.undp.org/content/dam/aplaws/publication/en/publications/environment-energy/www-ee-library/sustainable-energy/world-energy-assessment-energy-and-the-challenge-of-sustainability/World%20Energy%20 Assessment-2000.pdf.

In developing nations, discussing the energy demands of poor people, who show to a vastly larger part, will require major structural changes. Then again, in industrialized nations satisfactory access to moderate energy is difficult just for a minority, and hence amenable to social policy solutions. All through the world, nonetheless, poor family units pay a bigger portion of their income for energy than do the rich, as are powerless against fast increments in the cost of energy. Increments in the cost of oil in the winter of 1999/2000, for instance, represented a hardship for many people even in some industrialized nations (Goldemberg and Johansson, 1995; Fry, 1997).

Eradicating poverty is a long-term objective of improvement. However, sometime before that objective is accomplished, helpful, and affordable energy services could significantly improve expectations for everyday comforts and offer more opportunities to individuals. The present inequity is impractical. Fulfilling the energy needs of the poor with present-day advancements can improve ways of life and health, and make new openings and business opportunities (Runci, 2000a, b). Permitting 33% of the total populace to keep on persevering through the imperatives related to conventional vitality is unsuitable from a helpful and good viewpoint. Making business vitality all the more generally accessible bodes well from a humanitarian and moral point of view. The flood of democratization clearing the world is placing political force in the hands of the monetarily disenfranchised. Social orders with grave imbalances and aberrations will, in general, be shaky, and enormous populaces beneath the destitution line are fertile ground for social changes.

Table 3.2: Energy-Related Options to Address Social Issues

Social Challenge	Energy Linkages and Interventions
Alleviating poverty in developing countries	• Make commercial energy accessible to rise income-producing opportunities. • Increase productivity and improve health by giving universal access to suitable energy services—mainly for lighting, cooking, and transport—through high-quality, affordable, safe, and environmentally acceptable end-use devices and energy carriers.

Increasing opportunities for women	• Inspire the use of liquid or gaseous fuels and improved stoves to improve women's health and decrease indoor air pollution. • Use women's entrepreneurial and managerial skills to grow, run, and revenue from decentralized energy systems. • Support the use of affordable commercial energy to diminish time-consuming and arduous physical labor at work and home.
Speeding the demographic transition (too low mortality and low fertility)	• Decrease child mortality by presenting cleaner fuels and cooking devices providing potable, safe water. • Effect attitudes about opportunities for women and family size through communications made available through modern energy carriers. • Use energy initiatives to shift the comparative welfares and costs of fruitfulness—for instance, passable energy services can decrease the necessity for children's physical labor for household errands.
Mitigating the problems associated with rapid urbanization	• Decrease the 'push' factor in rural-urban immigration by refining the energy services in rural zones. • Deliver worldwide access to reasonable multimodal public transportation and carriage services. • Exploit the compensations of high-density settlements through land planning. • Take the benefit of new technologies to evade energy-intensive, environmentally unsound growth paths.

3.3. ENERGY, THE ENVIRONMENT, AND HEALTH

The natural effects of energy use are not new. For a considerable length of time, wood consumption has added to the deforestation of numerous territories. Indeed, even in the beginning times of industrialization, water, local air, and land contamination arrived at significant levels. What is generally new is an affirmation of energy linkages to provincial and worldwide natural issues and their suggestions. Even though energy's potential for improving human prosperity is verifiable, conventional energy production and utilization are firmly connected to natural degradation. This debasement undermines human wellbeing and personal satisfaction and influences biological diversity and ecological balance (Broadman, 1998; Bronicki, 2000).

The earth energy linkage is outlined in Table 3.3, which shows the portion of lethal emissions and different toxins inferable from the energy supply. The human disturbance record is the proportion of the human-created flow of a given toxin, (e.g., sulfur dioxide) to the common, or baseline, flow. Hence, on account of sulfur, the list is 2.7, which implies that human-created outflows of 84 million tons a year are 2.7 occasions the common baseline flow of 31 million tons per year. The table demonstrates that, together with other human exercises, energy systems fundamentally influence the worldwide cycling of significant synthetic concoctions. Even though without anyone else the record does not show that these emissions convert into negative effects, their extents give cautioning that such effects could be impressive. A few effects, as examined beneath, are as of now critical (Farinelli, 1999; Forsyth, 1999).

Table 3.3: Environmental Insults Due to Human Activities by Sector, the MID-1990s

Insult	**Natural Baseline (Tonnes per year)**	**Human Disruption Index**[a]	**Share of Human Disruption Caused by**			
			Commercial Energy Supply	**Traditional Energy Supply**	**Agriculture**	**Manufacturing, Other**
Sulphur emissions to atmosphere	31 million (sulphur)	2.7	85% (fossil fuel burning)	0.5% (traditional fuel burning)	1% (agricultural burning)	13% (smelting, refuse burning)
Oil added to oceans	200,000	10	44% (petroleum harvesting, processing, and transport)	Negligible	Negligible	56% (disposal of oil wastes, including motor oil changes)
Nitrous oxide flows to atmosphere	33 million	0.5	12% (fossil fuel burning)	8% (traditional fuel burning)	80% (fertilizer, land clearing, aquifer disruption)	Negligible
Lead emissions to atmosphere[b]	12,000	18	41% (fossil fuel burning, including additives)	Negligible	Negligible	59% (metal processing, manufacturing, refuse burning)
Cadmium emissions to atmosphere	1,400	5.4	13% (fossil fuel burning)	5% (traditional fuel burning)	12% (agricultural burning)	70% (metals processing, manufacturing, refuse burning)

Carbon dioxide flows to atmosphere	150 billion (carbon)	0.05e	75% (fossil fuel burning)	3% (net deforestation for fuelwood)	15% (net deforestation for land clearing)	7% (net deforestation for lumber, cement manufacturing)
Methane flow to atmosphere	160 million	2.3	18% (fossil fuel harvesting and processing)	5% (traditional fuel burning)	65% (rice paddies, domestic animals, land clearing)	12% (landfills)
Nitrogen fixation (as nitrogen oxide and ammonium)c	140 million (nitrogen)	1.5	30% (fossil fuel burning)	2% (traditional fuel burning)	67% (fertilizer, agricultural burning)	1% (refuse burning)
Particulate emissions to atmosphere	3,100 million	0.12	35% (fossil fuel burning)	10% (traditional fuel burning)	40% (agricultural burning)	15% (smelting, non-agricultural land clearing, refuse)
Mercury emissions to atmosphere	2,500	1.4	20% (fossil fuel burning)	1% (traditional fuel burning)	2% (agricultural burning)	77% (metals processing, manufacturing, refuse burning)
Non-methane hydrocarbon emissions to atmosphere	1,000 million	0.12	35% (fossil fuel processing and burning)	5% (traditional fuel burning)	40% (agricultural burning)	20% (non-agricultural land clearing, refuse burning)

Note: The greatness of the insult is just one factor deciding the size of the real ecological effect.

[a] *The human interruption index is the proportion of human-created stream to the natural (benchmark) stream.*

[b] *The automotive of human-actuated lead outflows right now thought to be 50% of worldwide automotive emissions in the mid-1990s.*

[c] *Determined from all-out nitrogen fixation short that from nitrous oxide.*

[d] *Dry mass.*

[e] *Although little, on account of the long air lifetime and different qualities of carbon dioxide, this slight unevenness in characteristic streams is causing a 0.4% yearly increment in the worldwide environmental centralization of carbon dioxide.*

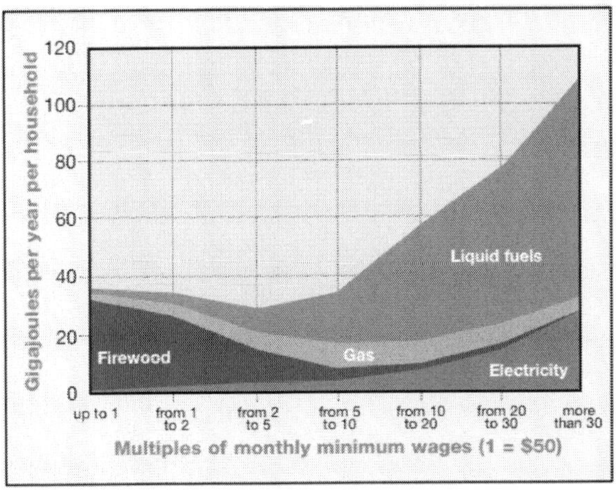

Figure 3.5: Average energy demand.

Source: https://www.undp.org/content/dam/aplaws/publication/en/publications/environment-energy/www-ee-library/sustainable-energy/world-energy-assessment-energy-and-the-challenge-of-sustainability/World%20Energy%20 Assessment-2000.pdf.

At each level (regional, local, worldwide), the ecological results of current examples of energy age and go through make a noteworthy division of human effects on the earth. At the family unit level, solid fuel use for cooking and warmth has noteworthy well-being impacts. Poor air quality—at the family unit, regional, and local levels—is related to expanded infection and sudden passing. Around 2 million unexpected losses a year—excessively of women and children—are assessed to happen from presentation to indoor air contamination brought about by consuming strong powers in inadequately ventilated spaces. Particulate issue (which is both radiated legitimately and shaped in the air as the aftereffect of the emanations of vaporous antecedents as oxides of sulfur and nitrogen) and hydrocarbons are developing concerns around the world. They are particularly troublesome in numerous pieces of the creating scene, where dirtier fills prevail with little emanations decrease. No protected limit level for presentation to the little particulate issue has been built up (Chantramonklasri, 1990; Bruce et al., 1996).

Combustion of fossil fuel is tricky on a few levels (although flammable gas produces fundamentally less destructive discharges than do oil or coal). The fundamental poisons radiated in the combustion of petroleum products are carbon monoxide, sulfur, and nitrogen oxides, and suspended particulate matter. Ozone is produced in the troposphere from interactions among nitrogen, hydrocarbons, sunlight, and oxides. Energy-related discharges from the combustion of fossil fuel, containing the transport sector, are significant supporters of urban air contamination (Runci, 1998, 1999). Precursors of acid deposition from fuel ignition can be precipitated a huge number of kilometers from their place of origin—often crossing national limits. The subsequent fermentation is making critical harm to natural systems, harvests, and human-made structures; and can, after some time, change the arrangement and function of whole biological systems. In numerous areas, acidification has lessened the profitability of woodlands, fisheries, and farmlands. Huge hydropower projects regularly raise ecological issues identified with flooding, though on account of atomic force, issues, for example, waste removal raise concern (Evans, 1999; Faninger, 2000).

The combustion of fossil fuel produces more carbon dioxide (CO_2) than some other human activity. This is the greatest source of the anthropogenic greenhouse gas discharge that is changing the organization of the air and could modify the worldwide climate system, including the pattern and amount of rainfall. Accomplishing a stable climatic CO_2 concentration at any level would necessitate that CO_2 emanations, in the long run, be cut by the greater part from current levels. Balancing out CO_2 at near the present concentration would require diminishing discharges to half of the current levels inside the following hardly any decades. Rather, CO_2 emanations keep on expanding. Current CO_2 expanding concentrations of greenhouse gases. The parity of proof, as indicated by the Intergovernmental Panel on Climate Change (IPCC), recommends that there is as of now a perceivable human effect on the global climate (Ruijgrok and Oosterhuis, 1997; Romm, 1999).

Since, by definition, sustainable energy systems must help both human and biological systems over the long term, objectives on tolerable discharges should look well into what's to come. They ought to likewise consider the open's tendency to request more health and natural security as flourishing increments.

Although the extent of ecological issues identified with energy may appear to be overpowering, various 'win-win' procedures could at the same time advantage the environment (at a few levels), the economy, and human prosperity. For instance, the substitute of solid fuels for cooking with liquid or gaseous fuels could have noteworthy ecological advantages at the regional, local, community, and worldwide scales, with attendant benefits for productivity and health (Engelman, 1998; Electric Power Research Institute, 1999).

3.4. ENERGY SECURITY

Energy security implies the accessibility of energy consistently in different forms, in adequate amounts, and at moderate costs. These conditions must prevail over a long period of energy is to add to economic advancement.

Consideration regarding energy security is important on account of the uneven distribution both of the capacity to develop other resources and of the fossil fuel resources on which most countries currently rely. The energy supply could turn out to be progressively vulnerable over the close to term because of the developing worldwide dependence on imported oil. For instance, the oil reliance (net imports as a portion of all-out interest) of OECD nations is required to develop from 56% in 1996 to 72% in 2010.

Moreover, although energy security has been satisfactory for as long as 20 years, and has in reality improved, the potential for strife, damage, interruption of exchange, and decrease in vital reserves cannot be expelled. These potential threats point to the need of fortifying worldwide just as local and national energy security (Piscitello and Bogach, 1998; Penner et al., 1999).

Choices to upgrade energy security include:

- Avoiding extreme reliance on imports by expanding end-user productivity and empowering more prominent dependence on nearby assets (especially those whose advancement will have other positive externalities, for example, work creation, capacity building, and contamination reduction), gave these do not include disproportionate expenses or waste rare assets.
- Diversifying inventory (counting the two providers and energy forms).
- Fostering more prominent political soundness through global cooperation and long term understandings among energy bringing in nations and among exporting and importing nations. Models may incorporate more extensive adoption—and progressively viable usage of—the Energy Charter Treaty, 11 just as expanded sharing of the foundation for moving natural gas.

- Encouraging innovation transfers (for instance, through joint endeavors and open private associations) to create nations so they can create nearby assets and improve energy efficiencies.
- Increasing national and provincial vital reserves of unrefined petroleum and oil items through expanded venture and propelled investigation advances.

Even though business sectors play a conspicuous role in making sure about energy supply in OECD nations, their role is humble in some developing nations and missing in others. Where markets do not thrive, the security of supply and services depends entirely on government activity and worldwide organizations, which may not serve the eventual benefits of customers. In such circumstances, energy security can be upgraded by empowering the improvement of structures that permit markets to add to the portion of energy assets (Eliasson, 1998; Emsperger, 2000).

As a result of little fuel necessities, nuclear power adds to the assorted variety of supply and supply security. Yet, public concerns about financial need, reactor safety, and radioactive waste transportation and removal—just as weapons multiplication—have controlled nuclear energy advancement in numerous nations. Anatomic mishap anyplace on the planet or a proliferation occurrence connected to nuclear power could additionally decrease support for nuclear power programs, with long term misfortune in the decent variety of the energy supply blend. Yet, on the off chance that by and largely acknowledged reactions could be found to the above concerns, nuclear energy could contribute essentially to make sure about power generation in numerous parts of the world (Eads and Peake, 1998).

People and business endeavors are likewise vulnerable to interruptions of energy supply. Even though the pattern towards the advancement of energy advertises by and large has improved energy security by offering more choices, supplies, and rivalry, it has additionally raised worries that the individuals who are ruined will be kept separate from the procedure, bringing about proceeded with energy weakness for certain people (European Commission, 1999).

3.5. ENERGY RESOURCES

Cautious examination of the long term accessibility of energy assets, beginning with unconventional and conventional oil and gas, demonstrates that these assets could last another 50–100 years—and potentially any longer—with known investigation and extraction innovations and foreseen specialized advancement in upstream activities. Coal assets and atomic materials are abundant to such an extent that they could, individually, keep going for quite a long time or centuries. Besides, although petroleum derivative costs may rise gradually after some time, the enormous, cost-driven increments in energy costs anticipated during the 1970s and 1980s will not happen within a reasonable time-frame (PCAST, 1997, 1999).

As proved by rising oil costs in the winter of 1999/2000, in any case, costs are dependent upon instability. This may happen, for instance, if cartels set costs autonomous of production costs. A few variances in costs can likewise be normal, particularly during the change to a huge scope utilization of unconventional oil and gas resources, because the planning of interests in upstream production limits may not compare with demand. Other cost-pushing elements could emerge from the ecologically all the more challenging extraction of unconventional oil assets (Dore and De Bauw, 1995; Duke and Kammen, 1999).

Renewable resources are more equitably conveyed than nuclear and fossil resources, and energy streams from sustainable resources are multiple sets of greatness higher than current worldwide energy use. Yet, the financial capability of renewables is influenced by numerous requirements—including contending land utilizes, the timing and amount of solar irradiation, natural concerns, and wind designs. Even though there are no genuine restrictions on future energy accessibility from a resource perspective, the presence of resources is of little pertinence without thought of how these can add to the stock of (downstream) energy administrations. Or maybe, the key concerns are: Can advancements to concentrate, gather, and convert these huge energy stocks and streams are created in time? Will these procedures have antagonistic ramifications? Will the energy benefits in the long run created from these resources be reasonable? Authentic proof recommends that these worries might be at any rate mostly balance by mechanical advancement, yet that such advancement should be supported—by guidelines to improve advertise execution, brief sponsorships, charge motivations, or different instruments—if it is to happen in an auspicious manner (Patterson, 1990, 1999).

3.6. ENERGY END-USE EFFICIENCY

The quadrupling of oil costs during the 1970s, the developing familiarity with energy-related contamination, and the nations could significantly benefit from the introduction of drastically new and increasingly proficient technologies in their energy-intensive fundamental materials processing. Since the possibility of environmental change has all added to a re-assessment of energy use. The outcome has been an improvement in the productivity with which energy is utilized in industry and power generation just as in lighting, household appliances, transportation, and warming and cooling of buildings (Dooley et al., 1998). This progressively effective utilization of energy is the main consideration adding to the enhancements in energy forces that have happened generally in practically all OECD nations, and all the more as of late in many change economies, just as in some in quickly developing nations, for example, Brazil, and China (O'Riordan and Jaeger, 1996; Obermair, 2000).

Today the worldwide energy effectiveness of changing over essential energy to valuable energy is around 33%. As it were, 66% of essential vitality is dispersed in the change forms, for the most part as low-temperature heat. Further huge losses happen when valuable energy conveys the energy administration. Various and different economic open doors exist for energy productivity upgrades, especially right now step from valuable energy to energy administrations. Making the most of these changes, which have gotten generally little consideration, has the biggest potential for cost-effective proficiency upgrades. It would mean less costly energy administrations and lower energy-related contamination and discharges (Dooley and Runci, 1999a, b).

Throughout the following 20 years, the quantity of primary energy required for a given degree of energy services could be cost-adequately decreased by 25–35% in industrialized nations (the higher figure being feasible by progressively viable approaches). These decreases are generally in the conversion step of valuable energy to energy benefits in the transportation, industrial, residential, commercial, and public sectors. Decreases of more than 40% are cost-effectively reachable on the move economies. Also, in most developing nations—which will, in general, have high financial development and old capital and vehicle stocks—the practical improvement possibilities run from 30 to more than 45%, comparative with energy efficiencies accomplished with existing capital stock (Dooley, 1999a, b, 2000).

The upgrades of around 2% a year suggested by the above figures could be improved by structural changes in industrialized and progress economies, by movements to less energy-intensive industrial production, and by immersion impacts in the transportation and residential areas. These consolidated impacts, made up by effectiveness enhancements and auxiliary changes, and could prompt reductions in energy intensity of 2.5% a year. The amount of this potential will be acknowledged relies upon the adequacy of strategy frameworks and measures, changes in perspectives and conduct, and the degree of a pioneering movement in vitality protection.

The following hardly any decades will probably observe new procedures, materials, motor systems, vehicles, and structures intended to lessen valuable energy these nations are as yet fabricating their physical foundation, they have a developing interest for basic materials. This opens a window of opportunities to advance and improve the efficiencies of production, especially in nations experiencing market change. The open doors are bigger at the purpose of a new venture, comparative with retrofitting (Dodds, 1997; Dooley, 1997).

Over the long term, extra, and dramatic gains in effectiveness are conceivable at all phases of energy transformation, especially from helpful energy to energy administrations. Examination shows that present advancements are not near arriving at hypothetical cutoff points and those enhancements of an order for magnitude for the entire energy framework may inevitably be accomplished (Cohen and Noll, 1991; Cooper, 2000).

For various reasons, the specialized and financial possibilities of energy productivity, just as its positive effect on sustainable advancement, have traditionally been under-figured it out. Accomplishing better quality use effectiveness includes an incredible assortment of specialized choices and players. Since it is a decentralized, scattered action, it is a troublesome issue for which to sort out help. Moreover, since it has little permeability, energy proficiency is not commonly a famous reason for lawmakers, the media, or people searching for acknowledgment and affirmation. Additionally, critical boundaries—basically showcase blemishes that could be overwhelmed by focused strategy instruments—forestall the acknowledgment of more noteworthy end-use efficiencies (Nye, 1990, 1998; Moor, 1997). These hindrances include:

- Lack of satisfactory data, technical information, and training.
- Uncertainties about the performance of investments in new and energy productive technologies.

- Lack of satisfactory capital or financing prospects.
- High perceived and initial expenses of increasingly proficient technologies.
- High transaction costs (for scanning and surveying data and training).
- Lack of incentives for cautious maintenance.
- The differential advantages to the client comparative with the investor (for instance, when energy bills are paid by the tenant instead of the landowner).
- External expenses of energy use, excluded from energy costs.
- Patterns and propensities for consumers, decision-makers, and operators, which might be affected by numerous variables, including thoughts of social glory and expert standards.

Acknowledging practical energy productivity possibilities will be helpful for singular energy consumers yet also to the economy in general. For instance, spared energy expenses can be utilized to produce energy sparing household products and services (Mitchell, 1998; Moisan, 1998). Also, as cost-effective energy upgrades are understood, extra beneficial open doors for development will keep on opening up because of innovative work, expectations to learn and adapt, and economies of scale. That implies that continual cost-effective energy productivity upgrades can be normal. Energy proficiency arrangements that utilization immediate or indirect value instruments, (e.g., the Renewable energy (RE) sources can give energy administrations zero or very nearly zero outflows of both air contaminations and green-house gases (Davis, 1995; DETR, 2000).

By and by, they can be significant energy sources in rural areas where network expansion is costly. They can likewise add to grid-connected power supplies in suitable hybrid arrangements. Discontinuous renewable can dependably give evacuation of subsidies and the consolidation of externalities) are viable in bringing down utilization trends in value delicate divisions and applications. However, even without changing the general value condition, energy proficiency approaches ought to be sought after to address market failures. For instance, effectiveness principles, machine, and item labeling, willful understandings, and professional training or contracting can build GDP development by improving natural and monetary execution, utilizing a given amount of vitality. Lawful principles; very much educated customers, organizers, and leaders; persuaded operators; and a sufficient installment framework for energy is fundamental to the fruitful usage of energy effectiveness enhancements (Darnell and Jefferson, 1994; Cowan, 1997).

3.7. RENEWABLE ENERGY (RE) TECHNOLOGIES

Sustainable power sources (counting biomass, wind, solar, hydropower, and geothermal) that utilizes indigenous assets can furnish energy services with zero or right around zero outflows of both air poisons and greenhouse gases. Right now, sustainable power sources supply 14% of the all-out world energy demand. The inventory is commanded by traditional biomass utilized for cooking and warming, particularly in the provincial areas of developing nations. Huge scope hydropower supplies 20% of worldwide power. Its degree for development is constrained in the industrialized world, where it has almost arrived at its financial limit. In the developing world, significant potential despite everything exists, except huge hydropower; undertakings may confront financial, ecological, and social constraints.

By and large, new sustainable power sources contributed 2% of the world's energy utilization in 1998, including 7 exajoules from present-day biomass and 2 exajoules for all different renewables (geothermal, wind, solar, and marine energy, and small-scale hydropower). Sun oriented photovoltaics and grid-connected wind introduced capacities are developing at a pace of 30% a year. All things being equal, it will probably be a very long time before these new renewables signify a significant part of all-out vitality utilization since they as of now speak to such a little rate (York, 1994).

Considerable value decreases in the previous decades have made some renewables competitive with non-renewable energy sources in specific applications in developing markets. Present-day, dispersed types of biomass appear to be especially encouraging for their capability to give rural areas clean types of energy-dependent on the utilization of biomass resources that have generally been utilized in wasteful, dirty ways. Biomass can be financially created with negligible or even positive ecological effects through perpetual yields. Wind power in waterfront and other breezy areas is promising as well (Mintzer and Leonard, 1994; Michaelis, 1996).

Not like conventional thermal power and hydropower sources, wind, and solar electric or thermal sources are intermittent.10–30% of absolute power supplies whenever worked related to hydropower-or fuel-based power generation. Developing stockpiling prospects and new techniques for working matrices offer a guarantee that the job of discontinuous advances could be significantly bigger (World Bank, 1998, 1999a, b, c).

Noteworthy hindrances, which could be overwhelmed by suitable systems and arrangements, hold up way of the quickened advancement of renewable innovations. These hindrances incorporate monetary dangers, administrative impediments, constrained accessibility of items, data, and innovation gaps, and absence of investment. The huge challenge is financial, even though expenses have descended fundamentally in recent decades. Table 3.4 outlines the status of different sustainable advances and gives data on patterns in cost and capacity (WRI, 1998; World Bank, 1999d).

Numerous sustainable advances, since they are little in scale and modular, are acceptable candidates for preceded with cost-cutting because of field understanding. The cost decreases of fabricated products, which are commonly fast from the outset and afterward decrease as the business develops, are called experience bends. These bends resulted in industry-wide cost decays of around 20% for each aggregate multiplying of production for solar photovoltaics, wind generators, and gas turbines because of learning impacts, peripheral mechanical upgrades, and economies of scale. Comparative decays are normal for other little scope renewables (Marshall, 1998). The quick development of renewable-based energy frameworks will expect activities to invigorate the market right now. This expansion can be accomplished by discovering approaches to drive down the general expense of renewables in their beginning periods of improvement and commercialization, while as yet exploiting the financial efficiencies of the commercial center. Valuing dependent on the full expenses of conventional energy sources (including eliminating subsidies and disguising externalities) will make renewables increasingly serious. Since disguising outside expenses might be dubious for quite a while, 'green' estimating of power and heat (which lets consumers pay more for ecologically kind energy supplies if they pick) might be a prompt choice in industrialized nations (Margolis and Kammen, 1999).

3.8. ADVANCED ENERGY TECHNOLOGIES

3.8.1. Fossil Energy

Sustainability objectives demonstrate the significance of advancing fossil energy innovations towards the long-term objective of almost zero air poison and greenhouse gas outflows without entangled finish of-pipe control advancements. Near-term advancements and procedures should bolster this long term objective (World Bank, 1994, 1996, 1997a, b, c, d).

The innovative insurgency under route in power generation, where advanced frameworks are supplanting steam turbine advances, supports this long-term objective. Petroleum gas-fired consolidated cycles offering low costs, high proficiency, and low ecological effects are being chosen any place natural gas is promptly accessible—in certain nations in any event, uprooting huge new hydropower projects. Cogeneration is more cost-effective and can assume a lot bigger job in the vitality economy—whenever dependent on gas turbines and consolidated cycles instead of on steam turbines (Maddison, 1995; Lutz, 1997).

Reciprocating motors and rising microturbine and energy unit advances are likewise strong candidates for cogeneration at littler scales, including apartments and commercial buildings. Coal gasification by partial oxidation with oxygen to create syngas (for the most part carbon monoxide and hydrogen) makes it conceivable to give power through integrated gasifier combined cycle (IGCC) plants with air contamination emanations about as low concerning natural gas joined cycles. Today power from IGCC cogeneration plants is frequently serious with power from coal steam-electric plants in either cogeneration or power configurations (Wallace, 1995, 1996; Watson et al., 1996).

Albeit synthetic fluid fuels made in single-item facilities are not competitive, superclean syngas-inferred engineered fuels, (e.g., synthetic center distillates and dimethyl ether) produced in polygeneration facilities that make a few items all the while may before long be. Syngas can be produced from gaseous petrol by steam improving or different methods or from coal by gasification utilizing oxygen, as noted. Extending markets for clean engineered fuels is probably going to come about because of toughening air contamination regulations. Synthetic fuels produced through polygeneration will be founded on flammable gas on the off chance that it is promptly accessible. Synthetic middle distillates so created are probably going to be competitive where ease petroleum gas is accessible (as at remote developing nation locales); the innovation may encourage exploitation of generally little remote gaseous petrol fields (Lovins, 1996; Luiten and Blok, 1999).

Inflammable gas-poor, coal-rich areas, polygeneration dependent on coal gasification is promising. Such frameworks may incorporate production of extra syngas for conveyance by pipelines to small-scale cogeneration frameworks in manufacturing plants and buildings—making conceivable spotless and effective utilization of coal at small as well as huge scales. Rapidly developing polygeneration action is now under the path in a few nations dependent on the gasification of low-quality oil feedstocks—a movement that is assisting with preparing for coal-based frameworks. Boundaries to widespread deployment of advanced polygeneration and cogeneration frameworks are primarily institutional. Most frameworks will create more power than can be expended, so accomplishing good financial aspects relies upon having the option to sell co-produced power at competitive costs into electric networks. Utility arrangements have frequently made doing so troublesome, yet under the competitive economic situations towards which electric frameworks are advancing in numerous areas, polygeneration, and cogeneration frameworks will regularly farewell (USDOE, 1998; UNDP, 1999).

Table 3.4: Current Status and Potential Future Costs of Renewable Energy Technologies

Technology	Increase in installed capacity in past five years (percent a year)	Operating capacity, end 1998	Capacity factor (percent)	Energy production, 1998	Turnkey investment costs (U.S. dollars per kilowatt)	Current energy cost	Potential future energy cost
Low-temperature solar heat	» 8	18 GWth (30 million m^2)	8–20	14 TWh (th)	500–1700	3–20 ¢/kWh	2 or 3–10 ¢/kWh
Wind electricity	» 30	10 GWe	20–30	18 TWh (e)	1100–1700	5–13 ¢/kWh	3–10 ¢/kWh
Solar photovoltaic electricity	» 30	500 MWe	8–20	0.5 TWh (e)	5000–10000	25–125 ¢/kWh	5 or 6–25 ¢/kWh
Solar thermal electricity	» 5	400 MWe	20–35	1 TWh (e)	3000–4000	12–18 ¢/kWh	4–10 ¢/kWh
Marine energy Tidal Wave Current OTEC	0 – – –	300 MWe exp. phase exp. phase exp. phase	20–30 20–35 25–35 70–80	0.6 TWh (e) Unclear Unclear Unclear	1700–2500 1500–3000 2000–3000 Unclear	8–15 ¢/kWh 8–20 ¢/kWh 8–15 ¢/kWh Unclear	8–15 ¢/kWh Unclear 5–7 ¢/kWh Unclear

Biomass energy Electricity Heata Ethanol	» 3 » 3 » 3	40 GWe > 200 GWth 18 billion liters	25–80 25–80	160 TWh (e) > 700 TWh (th) 420 PJ	900–3000 250–750	5–15 ¢/ kWh 1–5 ¢/ kWh 8–25 $/ GJ	4–10 ¢/ kWh 1–5 ¢/ kWh 6–10 $/ GJ
Geothermal energy Electricity Heat	» 4 » 6	8 GWe 11 GWth	45–90 20–70	46 TWh (e) 40 TWh (th)	800–3000 200–2000	2–10 ¢/ kWh 0.5–5¢/ kWh	1 or 2–8 ¢/kWh 0.5–5 ¢/ kWh
Hydroelectricity Large Small	» 2 » 3	640 GWe 23 GWe	35–60 20–70	2510 TWh (e) 90 TWh (e)	1000–3500 1200–3000	2–8 ¢/ kWh 4–10 ¢/ kWh	2–8 ¢/ kWh 3–10 ¢/ kWh

Note: The cost of grid-supplied electricity in urban areas ranges from 2–3 (c/kWh (off-peak) to 15–25 c/kWh) (peak).

The potential linkage among military and peaceful employments of atomic energy was perceived at the beginning of the atomic age. Endeavors to make a non-proliferation system through the Nuclear from proliferation inferred syngas. The effective advancement of fuel cells would encourage the introduction of H_2 for energy. Fuel cells are getting serious consideration, particularly for transportation, since they offer high effectiveness and close to zero air contamination emanations. Automakers are hustling to create fuel cell vehicles, with showcase passage focused for 2004–2010. The energy unit vehicle will seek the job of 'car of the future' with the half breed inner ignition motor/battery-powered vehicle previously being brought into the market (Loh, 1998; Lovei and Weiss, 1998).

Syngas-based power and H_2 production procedures likewise encourage detachment and storage of CO_2 from fossil energy frameworks, making it conceivable to acquire helpful energy with almost zero outflows of greenhouse gases without huge increments in energy costs. Late research proposes that the worldwide capacity with regards to making sure about the removal of CO_2 in topographical repositories may be satisfactory to discard CO_2 from non-renewable energy source use for many years, although more research is required to be sure about this (Leggett, 1999; Littlechild, 2000).

Other advanced innovations (ultra-supercritical steam plants, pressurized fluidized-bed ignition, coal IGCC dependent on fractional oxidation in the air for power generation, direct coal liquefaction for synthetic fuel generation) offer advantages comparative with conventional advancements. Be that as it may, not at all like syngas-based technologies, such alternatives in the close to term would not offer to make ways to the long term objective of almost zero emanations without noteworthy increments in costs for energy services (Thomas, 1992; UNCTAD, 1999) (Figure 3.6).

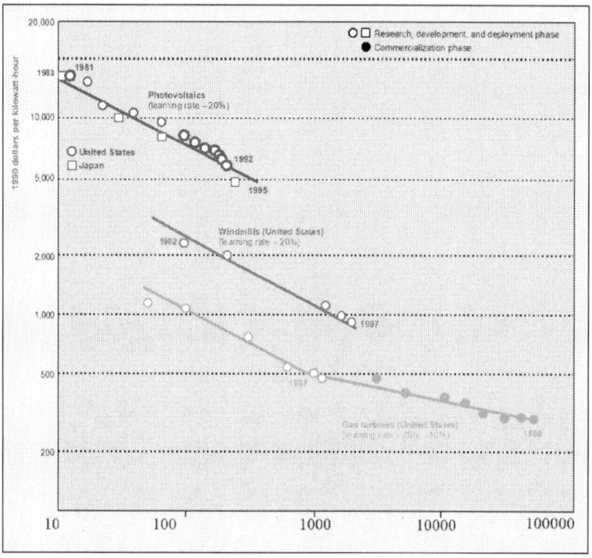

Figure 3.6: Technology performance and costs improve with experience, and there is a pattern of such improvements common to many technologies.

Note: The specific shape depends on the technology, but the persistent characteristic of diminishing costs is termed the 'learning' or 'experience' curve. The curve is likely to fall more sharply as technologies first seek a market niche, then full commercialization, because lower costs become increasingly important for wider success (Nakic'Enovic', Grübler, and McDonald, 1998).

Source: https://www.undp.org/content/dam/aplaws/publication/en/publications/environment-energy/www-ee-library/sustainable-energy/world-energy-assessment-energy-and-the-challenge-of-sustainability/World%20Energy%20Assessment-2000.pdf.

3.8.2. Nuclear Energy

Around the world, atomic energy represents 6% of energy and 16% of power. Albeit atomic energy rules power generation in certain nations, its underlying promise has not been generally figured it out. Most examiners project that atomic energy's commitment to worldwide energy will not develop—and may decrease during the underlying many years of the 21st century. Atomic power is more exorbitant than initially anticipated, competition from alternative innovations is expanding, and there has been lost in public certainty due to concerns identified with wellbeing, radioactive waste management, and potential atomic weapons expansion (Strickland and Sturm, 1998; Taylor and Bogach, 1998).

But since atomic power can give energy without transmitting ordinary air poisons and greenhouse gases, it merits investigating whenever advanced innovations could offer at the same time lower costs, help public trust in the safety of atomic reactors, guarantee that peaceful atomic projects are not utilized for military purposes, and exhibit effective atomic waste management rehearses. Not at all like Chernobyl-type reactors, have the light water reactors (LWRs) that overwhelm atomic power internationally had a decent safety record— even though this record has been accomplished at impressive expense to limit the danger of accidents (STAP and GEF, 1996; STAPPA and ALAPCO, 1999).

Non-Proliferation Treaty and a progression of local treaties, controls on trade in atomic materials and products and services that may be utilized to advance military desire, and defends applied to atomic materials in peaceful atomic applications have been to a great extent effective in isolating peaceful and military employments. If there is to be an energy future where atomic power, in the long run, contributes substantially more than at present, stronger institutional estimates will be expected to keep up this partition. These measures ought to be supplemented by technological advances planned for constraining chances to obtain atomic weapons under the appearance of peaceful atomic energy applications and to take weapons-usable atomic materials (Landes, 1969; Larsen and Shah, 1992, 1995).

Reactor improvement activity for the near term has included both evolutionary LWRs and new ideas. Reactor merchants presently offer a few evolutionary LWRs with improved security includes and standardized structures, for which there can be a high level of certainty that performance and cost targets will be met. Another developmental action includes modifying LWRs to make them more proliferation-resistant through a denatured uranium or thorium fuel cycle. One idea being returned to, the pebble bed modular reactor, offers the potential for a high level of inborn safety without the requirement for entangled, capital-intensive safety controls. A pebble bed modular reactor could likewise be worked on a proliferation-resistant denatured uranium-or thorium fuel cycle (Sørensen et al., 1999; Stahel, 2000).

Access to minimal effort uranium supplies could compel atomic power advancement dependent on LWRs. The plutonium breeder reactor, which requires reprocessing spent fuel to recuperate plutonium for reusing in fresh fuel, was once thought to be a suitable choice for addressing this challenge. Be that as it may, power costs for breeders would most likely be higher than for LWRs, in any event until late in the 21st century, and forestalling multiplication is considerably more testing with reprocessing and plutonium reusing than with LWRs worked on once-through fuel cycles (Jevons, 1965; Jefferson, 2000).

Other long-term choices for addressing to the atomic resource requirement are alternative breeder ideas—including atom smasher drove reactors, uranium from seawater, and thermonuclear fusion. The imminent costs, security, and proliferation obstruction highlights of such alternative breeder ideas are dubious, and the ideas would take a very long time to create. Late research recommends it may be plausible, at generally minimal effort, to separate uranium from seawater, where its concentration is low however, absolute amounts are immense. If the innovation could be conveyed at comprehensively huge scopes, it may be possible to abstain from making significant commitments to atomic fuel reprocessing and plutonium reusing. Fusion could give a practically boundless vitality supply, yet it will most likely not be economically accessible before 2050 (Schock et al., 1999; Sabugal Garcia, 2000).

Radioactive waste results of atomic energy must be confined so they can stay away forever to the human condition in concentrations that could cause huge harm. Even though the safety of long-term waste removal has not been demonstrated, the specialized network is certain that this goal can be acknowledged—to a great extent due to the little volumes of wastes included. In any case, in many nations, there is no social accord on the objectives and guidelines for radioactive waste removal and systems (both interim and long term) for actualizing them (Shapiro, 1989; Smallridge, 1999). The issues included are just incompletely specialized. The present social impasse on waste removal not just mists possibilities for atomic extension; it likewise has made spent fuel reprocessing a de-facto interim atomic waste management procedure in certain nations. This has happened even though fuel reprocessing does not offer monetary gains and does not take care of the waste removal issue—it simply buys time and is making enormous inventories of plutonium that must be discarded with low multiplication risk (Jefferson, 1998a, b, c).

REFERENCES

1. ADB (African Development Bank), (1999). *African Development Report (1999)* (Vol. 1, pp. 1–30). Oxford, Oxford University Press.
2. Adelman, M. A., (1972). *The World Petroleum Market* (Vol. 1, pp. 1–34). Baltimore, Md., Johns Hopkins University Press for Resources for the Future.
3. Adelman, M. A., (1995). *The Genie out of the Bottle, World Oil since 1970* (Vol. 1, pp. 1–44).Cambridge, Mass., MIT Press.
4. ASTAE (Asia Alternative Energy Program), (1998). *The World Bank Asia Alternative Energy Program (ASTAE)* (Vol. 6, pp. 1–37). World Bank, Washington, D.C.
5. Austin, D., (1998). Contributions to climate change, are conventional metrics misleading the debate? *Climate Notes* (Vol. 1, pp. 1–25). World Resources Institute, Washington, D.C.
6. Ballin, H. H., (1946). *The Organization of Electricity Supply in Great Britain* (Vol. 1, pp. 1–54). London, Electrical Press.
7. Baranzini, A., Goldemberg, J., & Speck, S., (2000). A future for carbon taxes. *Ecological Economics 32*(3), 395–412.
8. Barland, K., (1999). CO_2 capture and disposal. In: *Technology Assessment in Climate Change Mitigation, Report of the IPIECA Workshop* (Vol. 1, pp. 10–40). London, International Petroleum Industry Environmental Conservation Association.
9. Bowers, B., (1982). *A History of Electric Light and Power* (Vol. 1, pp. 1–54). London, Peter Peregrinus.
10. Boyle, G., (1996). *Renewable Energy, Power for a Sustainable Future* (Vol. 1, pp. 1–32). Oxford, Oxford University Press.
11. Brack, D., (1998). *Trade and Environment, Conflict or Compatibility?* (Vol. 1, pp. 10–29). London, Earthscan for the Royal Institute of International Affairs.
12. Brenton, T., (1994). *The Greening of Machiavelli, the Evolution of International Environmental Politics* (Vol. 1, pp. 1–22). London, Earthscan for the Royal Institute of International Affairs.
13. Broadman, H. G., ed., (1998). *Russian Trade Policy Reform for WTO Accession* (Vol. 401, pp. 1–50). World Bank Discussion Paper. Washington, D.C.
14. Bronicki, L. Y., (2000). Financing of private renewable energy projects.

In: Sayigh, A. A. M., (ed.), *World Renewable Energy Congress VI, Part III* (Vol. 6, pp. 1–60). Oxford, Elsevier.

15. Bruce, J. P., Lee, H., & Haites, E. F., (1996). *Climate Change (1995), Economic and Social Dimensions of Climate Change* (Vol. 1, pp. 1–30). Cambridge, Cambridge University Press for the Intergovernmental Panel on Climate Change.
16. Chantramonklasri, N., (1990). The development of technological and managerial capability in the developing countries. In: Chatterji, M., (ed.), *Technology Transfer in the Developing Countries* (Vol. 1, pp. 1–33). London, Macmillan.
17. Cohen, L. R., & Noll, R. G., (1991). *The Technology Pork Barrel* (Vol. 1, pp. 1–40). Brookings Institution, Washington, D.C.
18. Cooper, C., (2000). *Personal Communication* (Vol. 1, pp. 1–30). Director, South African Energy Association, Melville, South Africa.
19. Cowan, R. S., (1997). *A Social History of American Technology* (Vol. 1, pp. 1–44). New York, Oxford University Press.
20. Darnell, J. R., & Jefferson, M., (1994). *New Renewable Energy Resources, a Guide to the Future* (Vol. 1, pp. 1–30). London, Kogan Page.
21. Davis, M., (1995). *Institutional Frameworks for Electricity Supply to Rural Communities, a Literature Review* (Vol. 1, pp. 5–32). University of Cape Town, Energy and Development Research Centre, South Africa.
22. DETR (Department of Environment, Transport, and the Regions), (2000). *Climate Change, Draft UK Programme* (Vol. 1, pp. 5–29). London.
23. Dodds, F., (1997). *The Way Forward, Beyond Agenda 21* (Vol. 1, pp. 5–29). London, Earthscan.
24. Dooley, J. J., & Runci, P. J., (1999a). *Adopting a Long View to Energy R&D and Global Climate Change* (Vol. 1, pp. 5–29). PNNL-12115. Battelle Memorial Institute, Pacific Northwest National Laboratory, Richland, Wash.
25. Dooley, J. J., & Runci, P. J., (1999b). *Energy Research and Development in Germany* (Vol. 1, pp. 5–29). PNNL-12207. Battelle Memorial Institute, Pacific Northwest National Laboratory, Richland, Wash.
26. Dooley, J. J., (1997). *US National Investment in Energy R&D, 1974–1996* (Vol. 1, pp. 5–29). PNNL-11788. Battelle Memorial Institute,

Pacific Northwest National Laboratory, Richland, Wash.

27. Dooley, J. J., (1999a). *Energy Research and Development in Japan* (Vol. 1, pp. 5–29). PNNL-12214. Battelle Memorial Institute, Pacific Northwest National Laboratory, Richland, Wash.
28. Dooley, J. J., (1999b). *Energy Research and Development in the United States* (Vol. 1, pp. 5–29). PNNL-12188. Battelle Memorial Institute, Pacific Northwest National Laboratory, Richland, Wash.
29. Dooley, J. J., (2000). *A Short Primer on Collecting and Analyzing Energy R&D Statistics* (Vol. 1, pp. 5–29). PNNL-13158. Battelle Memorial Institute, Pacific Northwest National Laboratory, Richland, Wash.
30. Dooley, J. J., Runci, P. J., & Luiten, E. E. M., (1998). *Energy in the Industrialized World, Retrenchment and Refocusing* (Vol. 1, pp. 5–29). PNNL-12061. Battelle Memorial Institute, Pacific Northwest National Laboratory, Richland, Wash.
31. Dore, J., & De Bauw, R., (1995). *The Energy Charter Treaty, Origins, Aims, and Prospects* (Vol. 1, pp. 5–32). Royal Institute of International Affairs, London.
32. Duke, R. D., & Kammen, D. M., (1999). The economics of energy market transformation initiatives. *The Energy Journal, 20*(4), 15–64.
33. Eads, G. S., & Peake, S., (1998). *Global Transport and Energy Development, the Scope for Change* (Vol. 1, pp. 1–34). World Energy Council, London.
34. EC (European Commission), (1999). *Energy in Europe, Economic Foundations for Energy Policy* (Vol. 1, pp. 5–29). The Shared Analysis Project, Luxembourg.
35. Eliasson, B., (1998). The road to renewables, energy technologies for the post-Kyoto world. *World Energy Council Journal* (Vol. 1, pp. 38–44). London.
36. Emsperger, W., (2000). *Personal Communication* (Vol. 1, pp. 1–29). Siemens, Power Generation Division, Erlangen, Germany.
37. Engelman, R., (1998). *Profiles in Carbon, an Update on Population, Consumption and Carbon Dioxide Emissions* (Vol. 1, pp. 5–29). Population Action International, Washington, D.C.
38. EPRI (Electric Power Research Institute), (1999). *Electricity Technology Roadmap, Powering Progress-(1999) Summary and Synthesis* (Vol. 1, pp. 5–29). Palo Alto, Calif.

39. Evans, M., (1999). *Energy Research and Development in Italy* (Vol. 1, pp. 1–29). PNNL-13071. Battelle Memorial Institute, Pacific Northwest National Laboratory, Richland, Wash.
40. Faninger, G., (2000). *Personal Communication* (Vol. 1, pp. 1–34). University of Klagenfurt, Institute for Interdisciplinary Research and Continuing Education, Austria.
41. Farinelli, U., (1999). *Energy as a Tool for Sustainable Development for African, Caribbean and Pacific Countries* (Vol. 1, pp. 5–39). European Commission and UNDP. New York, United Nations.
42. Forsyth, T., (1999). *International Investment and Climate Change, Energy Technologies for Developing Countries* (Vol. 1, pp. 5–29). London, Royal Institute of International Affairs and Earthscan.
43. Fry, A., (1997). *Business and Climate Change, Case Studies in Greenhouse Gas Reduction* (Vol. 1, pp. 5–29). International Chamber of Commerce and World Business Council for Sustainable Development, Kyoto, Japan.
44. Goldemberg, J., & Johansson, T., (1995). *Energy as an Instrument for Socio-Economic Development* (Vol. 1, pp. 1–34). New York, UNDP.
45. Goldemberg, J., & Reid, W., (1998). *Issues and Options, the Clean Development Mechanism* (Vol. 1, pp. 5–29). New York, UNDP.
46. Goldemberg, J., & Reid, W., (1999). *Promoting Development while Limiting Greenhouse Gas Emissions, Trends and Baselines* (Vol. 1, pp. 5–29). New York, UNDP, and World Resources Institute.
47. Goldemberg, J., Johansson, T. B., Reddy, A. K. N., & Williams, R. H., (1988). *Energy for a Sustainable World* (Vol. 1, pp. 5–39). New Delhi, J. Wiley.
48. Grubb, M., Vrolijk, C., & Brack, D., (1999). *The Kyoto Protocol: A Guide and Assessment* (Vol. 1, pp. 2–29). Royal Institute of International Affairs, London.
49. Grübler, A., (1996). Time for a change, on the patterns of diffusion of innovation. *Daedalus, The Journal of the American Academy of Arts, 125*(3), 19–42.
50. Grübler, A., (1998). *Technology and Global Change.* Cambridge, Cambridge University Press.
51. Haegermark, H., (2000). *Personal Communication* (Vol. 1, pp. 1–34). Vice President, Elforsk, Stockholm, Sweden.
52. Hannah, L., (1979). *Electricity before Nationalization* (Vol. 1, pp.

5–29). London, Macmillan.

53. Hannah, L., (1982). *Engineers, Managers and Politicians* (Vol. 1, pp. 5–39). London, Macmillan.
54. Heede, R., Morgan, R., & Ridley, S., (1985). *The Hidden Costs of Energy, How Taxpayers Subsidize Energy Development* (Vol. 1, pp. 5–29). Center for Renewable Resources, Washington, D.C.
55. Hippel, E. V., (1988). *The Sources of Innovation.* Oxford, Oxford University Press.
56. Houghton, J. T., Jenkins, G. J., & Ephraums, J. J., (1990). *Climate Change: The IPCC Scientific Assessment* (Vol. 1, pp. 2–29). Cambridge, Cambridge University Press.
57. Hughes, T. P., (1983). *Networks of Power, Electrification in Western Society, 1880–1930* (Vol. 1, pp. 5–29). Baltimore, Md., Johns Hopkins University Press.
58. Hwang, R., (1995). *Money Down the Pipeline, Uncovering the Hidden Subsidies to the Oil Industry* (Vol. 1, pp. 5–49). Union of Concerned Scientists, Cambridge, Mass.
59. IEA (International Energy Agency), (1997a). *Energy Efficiency Initiative* (Vols. 1-Energy Policy Analysis and 2-Country Profiles and Case Studies, pp. 1–34). Paris.
60. IEA (International Energy Agency), (1997b). *IEA Energy Technology R&D Statistics, 1974–1995.* Paris.
61. IEA (International Energy Agency), (1997c). *The Link between Energy and Human Activity* (Vol. 1, pp. 5–29). Paris.
62. IEA (International Energy Agency), (1998). *World Energy Outlook* (Vol. 1, pp. 5–39). Paris.
63. Ito, F., (2000). *Personal Communication* (Vol. 1, pp. 5–29). Secretary General, Japan Energy Association, Tokyo.
64. Jefferson, M., (1983). Economic uncertainty and business decision-making. In: Wiseman, J., (ed.), *Beyond Positive Economics?* (Vol. 1, pp. 1–34). London, Macmillan.
65. Jefferson, M., (1996). *Climate Change (1995): The IPCC Second Assessment Report Reviewed* (Vol. 1, pp. 5–39). Report 5. World Energy Council, London.
66. Jefferson, M., (1998a). After Buenos Aires: A world looking to business and industry for help? *Geopolitics of Energy* (Vols. 1, 20, No. 11, pp.

5–29). Canada Energy Research Institute, Calgary, Alberta.

67. Jefferson, M., (1998b). *Instruments for Mitigating Climate Change* (Vol. 10, pp.1–20). World Energy Council, London.
68. Jefferson, M., (1998c). *The Kyoto Conference and Protocol: Report 8.* World Energy Council, London.
69. Jefferson, M., (2000). *Carbon Dioxide Emissions from (1990) to (1999): What's Been Happening? Ecoal.* (Vol. 35, pp. 1–30). World Coal Institute, London.
70. Jevons, W. S., (1965). *The Coal Question* (3rd edn.). New York.
71. Johnson, S. P., (1993). *The Earth Summit: The United Nations Conference on Environment and Development (UNCED)* (Vol. 1, pp. 2–29). London, Graham and Trotman.
72. Kelley, A. M., Jewkes, J., Sawers, D., & Stillerman, R., (1958). *The Sources of Invention* (Vol. 3, pp. 1–29). London, Macmillan.
73. Landes, D. S., (1969). *The Unbound Prometheus, Technological Change, and Industrial Development in Western Europe from 1750 to the Present* (Vol. 1, pp. 1–34). Cambridge, Cambridge University Press.
74. Landes, D. S., (1998). *The Wealth and Poverty of Nations: Why Some are so Rich and Some so Poor* (Vol. 1, pp. 5–29). London, Little, Brown.
75. Larsen, B., & Shah, A., (1992). *World Fossil Fuel Subsidies and Global Carbon Emissions* (Vol. 1, pp. 5–29). World Bank, Washington, D.C.
76. Larsen, B., & Shah, A., (1995). Global climate change, energy subsidies, and national carbon taxes. In: Bovenberg, L., & Cnossen, S., (eds.), *Public Economics and the Environment in an Imperfect World* (Vol. 1, pp. 5–29). Boston, Mass., Kluwer.
77. Leggett, J., (1999). *The Carbon War, Dispatches from the End of the Oil Century* (Vol. 1, pp. 5–29). London, Allen Lane/Penguin Press.
78. Littlechild, S. C., (2000). *Privatization, Competition and Regulation* (Vol. 1, pp. 5–22). Institute of Economic Affairs, London.
79. Loh, J., (1998). *Living Planet Report* (Vol. 1, pp. 5–44). Gland, Switzerland, World Wildlife Fund International.
80. Lovei, M., & Weiss, C., (1998). *Environmental Management and Institutions in OECD Countries, Lessons from Experience* (Vol. 1, pp. 5–29). World Bank Technical Paper 391. Washington, D.C.
81. Lovins, A. B., (1996). Negawatts, twelve transitions, eight

improvements and one distraction. *Energy Policy*.

82. Luiten, E. E., & Blok, K., (1999). *Energy Research and Development in the Netherlands* (Vol. 1, pp. 5–32). PNNL-12177. Battelle Memorial Institute, Pacific Northwest National Laboratory, Richland, Wash.
83. Lutz, W. L., (1997). *The Future Population of the World, What Can We Assume Today?* (Vol. 1, pp. 5–29) London, International Institute for Applied Systems Analysis and Earthscan.
84. Maddison, A., (1995). *Monitoring the World Economy 1820–(1992)* (Vol. 1, pp. 1–34). Paris, Organization for Economic Co-operation and Development.
85. Margolis, R. M., & Kammen, D. M., (1999). Underinvestment: The energy technology and R&D policy challenge. *Science, 285*, 690–92.
86. Marshall, L., (1998). *Economic Instruments and the Business Use of Energy* (Vol. 1, pp. 5–49). London, H.M. Treasury.
87. Michaelis, L., (1996). *Reforming Coal and Electricity Subsidies* (Vol. 1, pp. 5–29). Paris, Organization for Economic Co-operation and Development.
88. Mintzer, L. M., & Leonard, J. A., (1994). *Negotiating Climate Change: The Inside Story of the Rio Convention* (Vol. 1, pp. 5–31). Cambridge, Cambridge University Press.
89. Mitchell, C., (1998). *Renewable Energy in the U.K., Policies for the Future* (Vol. 1, pp. 5–29). Council for the Protection of Rural England, London.
90. Moisan, F., (1998). *Energy Efficiency Policies and Indicators* (Vol. 1, pp. 5–29). World Energy Council, London.
91. Moor, A. P. G. D., (1997). *Perverse Incentives* (Vol. 1, pp. 5–41). Earth Council, San José, Costa Rica.
92. Nye, D. E., (1990). *Electrifying America, Social Meanings of a New Technology, 1880–1940* (Vol. 1, pp. 5–29). Cambridge, Mass., MIT Press.
93. Nye, D. E., (1998). *Consuming Power: A Social History of American Energies* (Vol. 1, pp. 5–31).Cambridge, Mass., MIT Press.
94. O'Riordan, T., & Jaeger, J., (1996). *Politics of Climate Change: A European Perspective* (Vol. 1, pp. 5–29). London, Routledge.
95. Obermair, G., (2000). *Personal Communication* (Vol. 1, pp. 1–34). Secretary General, Austrian National Committee, World Energy

Council, Vienna, Austria.

96. OLADE (Organización Latinoamericana de Energia), (1997). *Energy and Sustainable Development in Latin America and the Caribbean* (Vol. 1, pp. 5–29). Quito, Ecuador.
97. Patterson, W. C., (1990). *The Energy Alternative* (Vol. 1, pp. 1–29). Boxtree/Channel 4, London.
98. Patterson, W. C., (1999). *Transforming Electricity: The Coming Generation of Change* (Vol. 1, pp. 5–29). London, Royal Institute of International Affairs and Earthscan.
99. PCAST (President's Committee of Advisors on Science and Technology), (1997). *Federal Energy Research and Development for the Challenges of the Twenty-First Century* (Vol. 1, pp. 1–29). Washington, D.C.
100. PCAST (President's Committee of Advisors on Science and Technology), (1999). *Powerful Partnerships: The Federal Role in International Cooperation on Energy Innovation* (Vol. 1, pp. 2–29). Washington, D.C.
101. Penner, J. E., Lister, D. H., Griggs, D. J., Dokken, D. J., & McFarland, M., (1999). *Aviation and the Global Atmosphere* (Vol. 1, pp. 5–21). A Special Report of IPCC Working Groups 1 and 3. Cambridge, Cambridge University Press for the Intergovernmental Panel on Climate Change.
102. Piscitello, E. S., & Bogach, V. S., (1998). *Financial Incentives for Renewable Energy Development* (Vol. 391, pp. 5–29). World Bank Discussion. Washington, D.C.
103. Romm, J. J., (1999). *Cool Companies: How the Best Businesses Boost Profits and Productivity by Cutting Greenhouse Gas Emissions* (Vol. 1, pp. 5–29). Washington, D.C., Island Press.
104. Ruijgrok, E., & Oosterhuis, F., (1997). *Energy Subsidies in Western Europe* (Vol. 1, pp. 5–29). Greenpeace International, Amsterdam.
105. Runci, P., (1998). *After Kyoto: Are There Rational Pathways to a Sustainable Global Energy System?* (Vol. 1, pp. 5–33). Paper presented at the Aspen Institute's Energy Policy Forum, Washington, D.C.
106. Runci, P., (1999). *Energy Research and Development in the European Union* (Vol. 1, pp. 1–29). PNNL-12218. Battelle Memorial Institute, Pacific Northwest National Laboratory, Richland, Wash.
107. Runci, P., (2000a). *Energy Research and Development in Canada*

(Vol. 1, pp. 3–29). PNNL- 13233. Battelle Memorial Institute, Pacific Northwest National Laboratory, Richland, Wash.

108. Runci, P., (2000b). *Energy Research and Development in the United Kingdom* (Vol. 1, pp. 5–23). PNNL-13234. Battelle Memorial Institute, Pacific Northwest National Laboratory, Richland, Wash.

109. Sabugal, G. S., (2000). *Personal Communication* (Vol. 1, pp. 1–34). Director, Engineering Research and Development, Grupo ENDESA, Madrid, Spain.

110. Schock, R. N., Fulkerson, W., Brown, M. L., San Martin, R. L., Greene, D. L., & Edmonds, J., (1999). How much is energy research and development worth as insurance? *Annual Review of Energy and the Environment, 24*, 487–512.

111. Shapiro, D. L., (1989). *Generating Failure, Public Power Policy in the Northwest* (Vol. 1, pp. 5–33). Lanham, Md., University Press of America

112. Smallridge, C. G., (1999). *Rural Energy and Development: A Study of Canadian Rural Electrification* (Vol. 1, pp. 2–39). C.G. Smallridge & Associates, Ottawa, Canada.

113. Sørensen, B., Kuemmel, B., & Meibom, P., (1999). *Long-Term Scenarios for Global Energy Demand and Supply* (Vol. 359, pp. 5–37). Roskilde Universitets Center, Denmark.

114. Stahel, C., (2000). *Personal Communication* (Vol. 1, pp. 5–39). Swiss Energy Council, Berne, Switzerland.

115. STAP (Scientific and Technical Advisory Panel) and GEF (Global Environment Facility), (1996). *International Industrial Collaboration for Accelerated Adoption of Environmentally Sound Energy Technologies in Developing Countries* (Vol. 1, pp. 5–35). Nairobi, Kenya.

116. STAPPA (State and Territorial Air Pollution Program Administrators) and ALAPCO, (1999). *Reducing Greenhouse Gases and Air Pollution: A Menu of Harmonized Options* (Vol. 1, pp. 5–39). Final Report. Washington, D.C.

117. Strickland, C., & Sturm, R., (1998). Energy efficiency in world bank power sector policy and lending: New opportunities. *Energy Policy, 26*(11), 873–84.

118. Taylor. R. P., & Bogach, V. S., (1998). *China: A Strategy for International Assistance to Accelerate Renewable Energy Development* (Vol. 388,

pp. 1–64). World Bank Discussion. Washington, D.C.

119. Thomas, C., (1992). *The Environment in International Relations* (Vol. 1, pp. 1–34). Royal Institute of International Affairs, London.
120. UNCTAD (United Nations Conference on Trade and Development), (1999). *World Investment Report (1999): Foreign Direct Investment and the Challenge of Development* (Vol. 1, pp. 1–39). New York and Geneva.
121. UNDP (United Nations Development Programme), (1999). *Human Development Report (1999)* (Vol. 1, pp. 4–34). New York.
122. USDOE (U.S. Department of Energy), (1998). *Technological Opportunities to Reduce U.S. Greenhouse Gas Emissions* (Vol. 1, pp. 5–39). A Report by National Laboratory Directors. Washington, D.C. Together with Appendix B, Technology Pathways Characterization—Working Document.
123. Wallace, D., (1995). *Environmental Policy and Industrial Innovation, Strategies in Europe, the US and Japan* (Vol. 1, pp. 1–34). London, Royal Institute of International Affairs and Earthscan.
124. Wallace, D., (1996). *Sustainable Industrialization* (Vol. 1, pp. 5–33). London, Royal Institute of International Affairs and Earthscan.
125. Watson, R. T., Zinyowera, M. C., & Moss, R. H., (1996). *Climate Change (1995), Impacts, Adaptation and Mitigation of Climate Change, Scientific-Technical Analysis* (Vol. 1, pp. 5–34). Cambridge, Cambridge University Press for the Intergovernmental Panel on Climate Change.
126. World Bank, (1994). *A Survey of Asia's Energy Prices* (Vol. 248, pp. 5–39). Technical Paper. Washington, D.C.
127. World Bank, (1996). *Rural Energy and Development, Improving Energy Supplies for Two Billion People* (Vol. 1, pp. 5–39). Washington, D.C.
128. World Bank, (1997a). *Clear Water, Blue Skies, China's Environment in the New Century* (Vol. 1, pp. 5–30). China 2020 series. Washington, D.C.
129. World Bank, (1997b). *Expanding the Measure of Wealth, Indicators of Environmentally Sustainable Development* (Vol. 1, pp. 2–39). Washington, D.C.
130. World Bank, (1997c). *World Development Indicators (1997)* (Vol. 1, pp. 5–39). Washington, D.C.
131. World Bank, (1997d). *World Development Report (1997): The State*

in a Changing World (Vol. 1, pp. 5–31). New York, Oxford University Press.

132. World Bank, (1998). *Financing Decentralized Renewable Energy* (Vol. 1, pp. 5–39). The Right Mix, Washington, D.C.
133. World Bank, (1999a). *Natural Gas, Private Sector Participation and Market Development* (Vol. 1, pp. 5–33). Washington, D.C.
134. World Bank, (1999b). *Global Development Finance, Analysis and Summary Tables* (Vol. 1, pp. 5–29). Washington, D.C.
135. World Bank, (1999c). *World Development Report (1998)/99, Knowledge for Development* (Vol. 1, pp. 5–59). New York, Oxford University Press.
136. World Bank, (1999d). *World Development Report (1999)/(2000), Entering the 21st Century* (Vol. 1, pp. 5–33). New York, Oxford University Press.
137. WRI (World Resources Institute), (1998). *World Resources: A Guide to the Global Environment, (1998)–(1999)* (Vol. 1, pp. 5–41). Washington, D.C.
138. York, D. W., (1994). Competitive electricity markets in practice, experience from Norway. *The Electricity Journal, 7*(5), 48–55.

CHAPTER 4

Opportunities and Issues in Renewable Energy Technologies for Sustainable Development

CONTENTS

4.1. Introduction 112

4.2. Socio-Cultural Barriers 112

4.3. Information and Awareness Hurdles 115

4.4. Market Fiascos and Economic Barriers 116

4.5. National and International and Strategies For Sustainable Development 119

4.6. Private, Local, and Nongovernmental Sustainable Development (SD) Initiatives 125

References 127

4.1. INTRODUCTION

Trailing any strategy of renewable energy (RE) deployment in the perspective of sustainable development (SD) entails that all social, environmental, as well as economic impacts are explicitly taken into consideration. Policy, integrated planning, and implementation methods can aid this through anticipating and overpowering prospective hurdles as well as manipulating opportunities of RE deployment. In this chapter, hurdles particularly associated with SD have been discussed. Such hurdles may either obstruct RE deployment or else consequence in trade-offs with SD criteria. In the opportunities section, the focus has been put on the integration of RE policies and measures in strategies for SD at different levels that can help in overcoming such hurdles and creating opportunities for RE deployment which can more entirely meet goals of SD (Wolsink, 2000, 2007b, 2010).

In this manner, RE policies as well as planning of project, its construction, and operation are embedded in the particular economic, social, and environmental framework and support the strategic development goals of a particular society. Another important thing is that they should stay aligned with multidimensional environmental agreements as well. In this chapter, in addition to looking at few major socio-cultural, and certain economic hurdles to RE deployment, links have been provided for each category of barriers, and hence should be taken into consideration during RE policy development and deployment (Devine-Wright, 2005, 2009).

4.2. SOCIO-CULTURAL BARRIERS

Majority of communities have conventionally regarded RE applications as eco-friendly thereby a high level of public support for RE has been documented in available literature and opinion polls (Devine-Wright, 2005; McGowan and Sauter, 2005; Wolsink, 2007b). Nevertheless, public support for RE at such generic level does not essentially indicate an active support and approval of RE at the local implementation level since in that case, RE deployment is frequently connected with direct impacts for groups as well as individuals (Painuly, 2001; Bell et al., 2005; Wustenhagen et al., 2007). For instance, in numerous countries, increased public resistance has been observed to new, large, installations, and often such resistance is more than the narrow 'not in my backyard' type of resistance (Wolsink, 2007b).

Regarding the deployment of RE and its probable SD trade-offs, there are different origins of socio-cultural barriers, and are intrinsically connected to personal norms as well as societal values (Sovacool and Hirsh, 2009). Such norms and values may have impact on the perception and acceptance of RE technologies and the prospective effects of their deployment by groups, individuals, and societies (GNESD, 2007b; Sovacool, 2009; West et al., 2010). From SD perspective, barriers are likely to rise from insufficient attention to such socio-cultural issues, which include hurdles associated with behavior; natural habitats and human and natural heritage sites, as well as effects on ecosystems and biodiversity; landscape aesthetics; and land / water use rights as well as their accessibility for challenging uses. These hurdles have been briefly discussed here (GNSED, 2004, 2007a).

Deployment of RE technologies can be connected with *behavioral* implications which challenge cultural as well as social norms, values, and perceptions (Reddy and Painuly, 2004; GNESD, 2007b; Chaurey and Kandpal, 2010). For instance, in India, multi-criteria analysis of home cooking devices shows that the second most significant thing while determining consumer preferences for home cooking devices are behavioral concerns and they are only surpassed by technical criteria (Pohekar and Ramachandran, 2006). Due to behavioral concerns, uptake is limited not just of the comparatively novel and technically advanced solar cookers. It also suggests a significant explanation for the non-usage of installed better-quality fuelwood cook-stoves in majority of India, where merely 6 million out of a total of 23 million installed better-quality fuelwood stoves were observed to be functional (Neudoerffer et al., 2001). Likewise, in several other developing countries, similar findings have been found regarding the importance of behavioral barriers for dissemination and usage of better quality cooking-stoves (Ben Hagan, 2003; Zuk et al., 2007; Bailis et al., 2005, 2009). Such behavioral barriers to new RE systems and technologies might be comparatively small given that the transition tries to emulate prevailing practices and properties of present technologies. Nonetheless, they tend to rise with the degree of changes in or consumption behavior levels (Kumar et al., 2009; Petersen and Andersen, 2009).

We do not always apply the precautionary principle for minimizing impacts on *human heritage sites* and *natural habitats* (Rylands and Brandon, 2005; Hreinsson, 2007; Nandy et al., 2007; Clarke, 2009; Hennenberg et al., 2010). This has resulted in public resistance to numerous types of RE development projects. General public perception of effects associated with aesthetics of transformed landscapes due to developments in wind power in OECD countries is a major hurdle that has been extensively studies in the literature (Wolsink, 2000, 2007b, 2010; Upreti, 2004; Jobert et al.,

2007; Wustenhagen et al., 2007). Public attitudes to offshore wind farms noticeable from shore is dependent on, for instance, the frequency and type of beach use with consistent visitors who consider coastal landscapes as more immaculate resources and hence less suitable for industrial usage (GNESD, 2008, 2010; Ladenburg, 2010).

Displacement and relocation of communities in project developments involving huge area of land like large-scale hydropower might be substantial. It has been estimated by The World Commission on Dams (2000) that globally, 40 to 80 million people have been moved owing to large dams. Moreover, this figure increases considerably when we take into account the concomitant impacts of alterations in freshwater ecosystems and river flows on downstream populations (Richter et al., 2010). Even though latest figures on the amount of people who are affected by developments in hydropower technology are available both at the country level as well as at the individual project level, cumulative statistics appear to be limited to the 2000 as per the report published by the World Commission on Dams. Additionally, often large-scale hydropower projects are linked with trade-offs associated with competing uses of water, for instance, for water supply for industrial and domestic purposes, irrigation, and flood control. Relocation of populations who are affected through developments in large-scale hydropower projects is obviously connected to the issue of land use rights of native people and associated with complicated relocation and compensation issues (Moore et al., 2010; Ölz and Beerepoot, 2010; Chen, 2009; Mirza et al., 2009). For instance, government might offer insufficient economic compensation to affected populations or to such people who are affected by externalities like losses in cultural heritage (Cernea, 1997; World Commission on Dams, 2000; Bao, 2010; Brown and Xu, 2010). Another area getting increasing attention is land usage concerns rising from commercial-scale energy crops (IIED, 2009). In this context, it is imperative to consider occupational concerns regarding labor and human rights, for example, working conditions in field crop projects (ILO, 2010). Lastly, another significant social concern is food security to which certification schemes are now paying more attention. As already mentioned, public acceptance and awareness is a significant element which must be dealt appropriately for speedily and significantly scaling-up RE deployment in order to help achieve climate change mitigation objectives. We can only undertake large scale project implementations through the support and understanding of the public (Bailis and Cutler, 2004; Zoellner et al., 2008). For this purpose, dedicated communication efforts are required related to the opportunities as well as achievements connected to the large-scale applications (Barry et al., 2008). Moreover, at the same time, participation of public in decisions related to planning as well as equity and fairness considerations in the distribution of the costs and benefits of RE deployment play a similarly significant role and cannot be ignored (Malesios and Arabatzis, 2010).

4.3. INFORMATION AND AWARENESS HURDLES

A common argument often quoted for promoting RE projects is that they contribute in reducing poverty as local communities benefit from increased employment opportunities, investment opportunities, skills development, and technology transfer (Goldemberg and Coelho, 2004; Modi et al., 2006; Goldemberg et al., 2008; Barbier, 2009). In developing countries, several RE pilot projects give subjective evidence of the part played by renewable sources in energy-poor communities (Karekezi and Kithyoma, 2003; Mondal et al., 2010). Nonetheless, if the local community fails to realize these benefits, or their distribution is reflected as unbalanced, acceptance of project might be difficult (Gunawardena, 2010). In unindustrialized countries, inadequate business and technical skills as well as lack of technical support systems are mainly present in the energy sector, where awareness as well as dissemination of information about existing suitable RE options among prospective consumers is a determining factor for uptake and market creation. Frequently, this gap in awareness is considered as the single most significant factor distressing the deployment of RE as well as development of both small and medium enterprises which contribute greatly to the economic growth. Disregarding the informational and perception issues related to the decentralized units can regularly lead to dysfunctional or abandoned systems (Werner and Schaefer, 2007).

In scenarios where private sector companies have proprietary ownership of RE technology and the diffusion of technologies also generally takes place through such markets where companies are key actors, it is essential to pay attention to the capacity of these companies to develop, implement, as well as deploy RE technologies (Wilkins, 2002). Hence, the significance of increasing business and technical competence as an integral part of capacity building—at both the micro-level as well as at the firm level—must be addressed (Lall, 2002; Figueiredo, 2003).

Attitudes towards RE are determined by more than just facts and knowledge. It is important to consider values and norms, and might impact personal and public perceptions of the effects of RE for consumption in addition to deeply held values such as control, trust, and freedom (Walker et al., 2010). This indicates that attitudes towards RE along with rationality are driven by emotions as well as psychological issues (Bang et al., 2000). To be successful, RE deployment and awareness efforts and approaches must take this explicitly into consideration particularly as hurdles to awareness and information may have implications for RE uptake, markets, uncertainty, and therefore capital expenditure (Jager, 2006; Nannen and van den Bergh, 2010; Litvine and Wüstenhagen, 2011).

4.4. MARKET FIASCOS AND ECONOMIC BARRIERS

In the context of SD, social costs, as well as benefits, must be taken into consideration to assess the economics of RE. RE must be evaluated against quantifiable criteria directed at regional suitability, cost-effectiveness, as well as distributional and environmental consequences (Gross, 2007; Creutzig and He, 2009). From the social point of view, a level economic playing field is essential for supporting rational RE investment decisions. This indicates that market distortions like subsidies and taxes and their structure, as well as market failures and imperfections, should be carefully considered with regard to their effects for RE deployment and the internalization of social expenditure, such as damages from GHG emissions, and environmental expenses (Rao and Kishore, 2010).

The size of the grid and technologies are the main factors of the *economic sustainability* of RE and of the effectiveness of RE as compared to non-renewable energy techniques. Suitable RE technologies that are economically sustainable are every so often observed to be available for increasing rural off-grid energy access (Bishop and Amaratunga, 2008; Ravindranath and Balachandra, 2009; Thompson and Duggirala, 2009; Deichmann et al., 2011). In the case of smaller off-grid applications, proof exists that numerous RE technologies, including mini-hydro, wind, and biomass-electric, can provide the lowest levelized generation expenditures of electrification, that is, including the levelized expenses of distribution and transmission. A number of RE technologies such as biomass, wind, geothermal, and hydro, are also possibly the least-cost mini-grid generation technologies (ESMAP, 2007). Nonetheless, non-renewable power generation technologies are still more sustainable economically as compared to RE in various circumstances (van Alphen et al., 2007; Cowan et al., 2009). This is predominantly the case for majority of large grid-connected applications, even with rise in oil rate forecasts and when prospective RE technology price reductions over the subsequent 20 years are taken into consideration.

Assessments of the economic sustainability of RE are centered on and subject to expectations regarding the expenditure and availability of the renewable resource. The absence of sufficient resource potential data impacts uncertainty about resource availability in a direct manner, which might turn into higher risk payments by project developers and investors, as seems to be the situation with geothermal electricity production in Indonesia. An evolving area seeking attention is related to the prospective effects of climate change and climate variability on energy resources and services, where the availability and timing of RE resources are directly affected (World Bank, 2011). Numerous impacts due to climate change and extreme events (for instance, typhoons, and hurricanes, floods, heatwaves, and famines) on energy resources and services have already been being experienced. For example, in Eastern Africa, the availability of power supply is profoundly dependent on hydropower, thereby, as a result of recent droughts, an annual costs of the order of 1 to 3.3% of annual GDP was estimated for concerns associated with the higher costs of RE owing to their variable availability (Eberhard et al., 2008; Karekezi et al., 2009).

In certain situations where the deployment of RE is practical from an economic point of view, several other financial and economic barriers might play a role in affecting the deployment of RE. Few examples of normally identified obstructions to RE deployment include high upfront investments expenses, including high grid connection and installation expenditures (Limmeechokchai and Chawana, 2007; Kassenga, 2008; Mathews, 2008; Monroy and Hernandez, 2008; Green and Vasilakos, 2011). In low-income countries, mostly, high upfront expenses of RE technologies may play a significant role in inhibiting uptake by consumers. Consumers might choose to keep the preliminary cost low instead of minimizing the operating expenses, which run through an extensive-time period, or they might have no option if they do not have enough cash or credit. Henceforth, to some extent, the effective uptake of RE technologies is dependent on the selection as well as the set-up of the dissemination model like cash sales, donations, consumer credits, or fee-for-service systems (Nieuwenhout et al., 2000).

For stimulation of economic growth and SD as well as for catalyzing rural and urban cash economies, both policy and entrepreneurial support systems are required in addition to the deployment of RE (Davidson et al., 2003). For instance, investments are needed for ensuring disposal of the technical capability needed for requisite operation and maintenance of the systems; however, investments are a major barrier for harnessing available RE sources in developing countries. A novel set of thinking is increasingly developing as well, where RE is treated as a vital part of a market-based energy economy and hence more sturdily including the private sector (GNESD, 2007b, 2008).

High upfront expenses may possibly also reveal high-risk perceptions of stakeholders as well as a general deficiency of financing instruments and underdeveloped or split financial sectors (Brunnschweiler, 2010). In this manner, subjective data from South East Asia proposes that insufficient experience and lack of understanding of RE systems amid financial institutions and stakeholders results in low participation by national financiers, which in turn may result in raising the expenditure of investment for RE projects through higher risk premiums. For example, in Indonesia, power projects based on biomass have been noticed as facing extra hurdles associated with an overall lack of knowledge in bioenergy project development and linked feedstock supply concerns among national stakeholders and banks.

The impacts of the timing of the torrent of expenditures and profits from RE investments results in a trade-off with regard to sustainability, for instance, in circumstances where decision-makers in unindustrialized countries have to elect amid investments in non-renewable energy associated with smaller payback time, nevertheless higher external expenditures, and RE investments with extended payback time, but higher positive externalities such as health, job creation, drop-in GHG emission, etc.

As a result of market distortions, different externalities takes place and are central when we address RE deployment in the context of SD. For instance, the structure of taxes and subsidies may favor non-renewable energy with hostile consequences for the competitiveness of RE. Likewise, current engineering capacities and grid networks will benefit some forms of energy above others, with repercussions for the path dependence of energy deployment. Such path dependencies can engage societies into electing such infrastructure or energy options which might be low-grade with regard to cost efficiency or accumulated social outlays in the long run (Unruh, 2000). In numerous situations, internalization of environmental externalities results in significant implications for the levelized outlays of RE technologies and afterward their non-inclusion causes an obstruction for deployment of RE

(Harmon and Cowan, 2009; Fahlen and Ahlgren, 2010). Take an example of internalization of damage costs which results from burning of fossil fuels which otherwise could result in a number of financially competitive RE technologies (Owen, 2006). Analogous conclusions were reached upon for PV mini-grids for three distant rural areas in Senegal, where levelized electricity expenses from PV technologies were observed to be lesser than the expenses of energy resulting from grid extension when different environmental externalities are taken into consideration (Thiam, 2010; Evans et al., 2009; Brent and Rogers, 2010).

Several modern studies include a number of environmental and social sustainability indicators for assessment as well as ranking of energy options. Besides GHG emissions, these sustainability indicators take account of consumption of water, land requirements, social effects, as well as availability of renewable sources thereby providing further insight into prospective barriers for RE deployment in the context of SD (Afgan et al., 2007; Becerra-Lopez and Golding, 2008; Brent and Kruger, 2009).

4.5. NATIONAL AND INTERNATIONAL AND STRATEGIES FOR SUSTAINABLE DEVELOPMENT

Different strategies for SD at national, international, and local levels as well as in government and private circles of society can aid us in overcoming barriers and creating new opportunities for RE deployment by integration of practices and policies of RE and SD. Different national and international levels strategies comprise: removal of such practices which work against SD; adopting such practices for SD that internalize social as well as environmental externalities; and integrating SD and RE strategies. At the indigenous level, SD initiatives taken by local governments, cities, as well as government and private organizations can be change drivers and play a part in overcoming common resistance to RE installations.

Since the Stockholm Conference on the Human Environment in 1972, the requirement of developing cross-sectoral, SD strategies have been expressed at the multilateral levels (Founex Committee, 1971; Engfeldt, 2009). The apprehensions were covered in the objectives of Agenda 21 which focused on adopting such strategies to harmonize various sectoral processes (UNCED, 1992; Steurer and Martinuzzi, 2007). In 2002, Johannesburg Plan of Implementation was approved at the World Summit on SD where several governments were called upon with a sense of earnestness in order to significantly upturn the worldwide share of RE as well as to take quick steps towards developing National sustainable development strategies (NSDS) by 2005 (UN, 2002). While formulating such NSDS, majority of countries generally gave preference to strategic policy areas and tangible goals for which both international commitments as well as national circumstances needed swift action like controlling climate change and growing use of RE (OECD, 2002; UNDESA, 2008). This type of prioritization can result in increased productivity, education, and health, growth of income, gender equality, human development, decreased social impacts related to energy extraction, as well as macroeconomic stability and control (World Bank, 2001). Particularly, RE technologies can add several other advantages. Moreover, integration of RE policy with NSDS delivers an outline for countries using which they can opt explicit policy instruments for incorporating apprehensions of other countries into their own, and aligning with global policy measures.

4.5.1. Elimination of Mechanisms Working Against Sustainable Development (SD)

By removing fossil fuel subsidies, government can open up several opportunities for more widespread usage or even market entry of RE. By doing so, artificially amplified competitive lead of fossil fuels will be decreased; hence, fossil fuel subsidies might instead be used for research and development as well as deployment of RE technologies. A report was prepared for the successive G-20 Summit by the OECD, IEA, and World Bank (2010) according to which the government aid of fossil fuels is geographically concentrated. In year 2009, thirty-seven economies, mostly non-OECD, accounted for above than 95% of fossil fuel subsidies globally demonstrating a total worth of USD_{2005} 268 billion. The support of government regarding fossil fuels is largest in such economies where there are ample supported energy carriers, for instance, Saudi Arabia and Iran. Fuels which are mostly supported are natural gas (USD_{2005} 73 billion) and oil (USD_{2005} 108 billion) and might also indirectly cover electricity (USD_{2005} 82 billion), if it is mainly generated by these fuels. Contrarily, international coal subsidies are relatively smaller at just USD_{2005} 5 billion.

A common concern while reforming these subsidies is their impact on the poor; hence, they should be designed cautiously as families with low-income are expected to be affected disproportionally. Nonetheless, quite often, subsidies are regressive and there is a considerable profit outflow to higher income groups (Del Granado et al., 2010). For instance, in Iran, the richest 30% people take advantage of 70% of whole government support, whereas, in Indonesia, the bottom 40% people belonging to low-income families reap merely 15% of entire energy subsidies. Generally, this takes into account most supported fuels, for example, LPG in India (Gangopadhyay et al., 2005), electricity in numerous African countries (Angel-Urdinola and Wodon, 2007), and petroleum products universally (Coady et al., 2010). Nonetheless, the picture is less clear in case of kerosene, and subsidies are comparatively better aimed (Coady et al., 2004).

For that reason, reforming subsidies in a way to increase the use of RE technologies must go together with addressing the particular needs of the poor. For doing so, there are two suitable general directions. The first course is increasing rural electrification, as underprivileged households have a tendency to live in areas where there is no electricity service. Several successful programs have been introduced in Vietnam and Ethiopia (IEA/OECD/World Bank, 2010), and the phase-out of parallel fossil fuel subsidies might generate added incentives for generating business activities (Barnes and Halpern, 2001). Expanding electrification can be accompanied with further support for RE technologies in centralized power supplies, which then will also be available to the poor. Secondly, in case electrification is not feasible or there are any better low-cost options, an alternative option is RE off-grid technologies. For instance, in Nepal, financial supports have considerably improved the awareness levels in adopting RE off-grid technologies as well as the will to pay for electricity (Mainali and Silveira, 2011). Likewise, in India, for home lighting, modern bioenergy systems and solar photovoltaics have become popular options in rural areas as compared to conventional kerosene-based lighting (Mahapatra et al., 2009).

There are possibly many more such opportunities; nevertheless further case-specific analysis is required for identification of potential gains for RE and evaluating its efficiency. Without this type of analysis it is not clear that RE technologies profit from phase-out of fossil fuel subsidies in a direct manner; moreover, this analysis determines whether the phase-out as such is possibly harmful or not.

In numerous event and studies, the significance of reducing barriers to trade in RE supplies and its related technologies has been stressed as part of a comprehensive strategy for decreasing reliance on less secure and more-polluting energy sources. This is the case for wind turbines, PV, and biofuels (Steenblik, 2005; Lucon and Rei, 2006; Tamiotti et al., 2009). Obstructions in the market penetration and global trade of bioenergy are not limited to but include tariff barriers, inaptly constricting sustainability benchmarks, and certification systems for biofuels and biomass, technical standards, logistical obstructions, and sanitary requirements. In general, by reducing or altogether eliminating barriers to trade, we can facilitate access to RE as well as other environmental goods which can significantly add to climate change mitigation by way of adopting better distribution of technologies at lower rates. Eradication of both tariffs as well as non-tariff barriers to clean technologies can possibly lead to a 14% rise in trade of these products (Brewer, 2004; WTO, 2010).

4.5.2. Mechanisms for Sustainable Development (SD) Which Internalize Social and Environmental Externalities

There is a continuous requirement for development of such mechanisms for SD which internalize social as well as environmental externalities. Dissemination of RE technologies is driven by such incentives and policies which assist in overcoming high upfront expenditures. Still, when we include external costs, the comparative benefits of renewable energies are highlighted—particularly concerning GHG emissions (Onat and Bayar, 2010; Varun et al., 2010). For incorporation of external costs, good indicators are required. A methodological restraint observed while studying different energy production systems is that they use an inadequate number of comparable sustainability indicators, which can result in flaws and biases in the ranking of energy technologies and sources against sustainability (Eason et al., 2009; Kowalski et al., 2009). Even though a significant contribution is made by multi-criteria decision analysis and tactics, it has been observed that evaluating the contribution of RE options to SD is not a straightforward task, taking into account different aspects of SD, the ambiguity of the associated information as well as the exemplified qualitative aspects which cannot be represented only by numerical values (Cavallaro, 2009; Michalena et al., 2009; Donat Castello et al., 2010).

A practical example of a mechanism for SD is the CDM established under the Kyoto Protocol. Under the CDM, it was expected that by 2012, RE as an alternative for fossil fuels would set up 61% of projects and 35% of anticipated Certified Emission Reductions. The CDM is extensively regarded as one of the best innovative features of the Kyoto Protocol involving 69 developing countries for creating an international carbon market worth several billion US dollars. Nevertheless, it is also commonly acknowledged that its contribution to low-carbon and SD paths in most host countries is dubious (Figueres and Streck, 2009). Designated National Authority (DNA) submits CDM projects for sustainability screening and approval. Still, there is no universal standard for sustainability assessment to deal with weaknesses in the prevailing system of sustainability approval (Olsen and Fenhann, 2008). Therefore, Designated National Authorities have a significant role in meeting national SD priorities as well as in attracting ventures (Winkler et al., 2005). By reviewing literature of the CDM (Paulsson, 2009) as well as its contribution to SD, we observe that one of the key weaknesses of the market mechanism is that preference is given to cheap emission reduction projects over more expensive projects which every so often are associated with higher SD benefits (Sutter and Parreño, 2007). There are different voluntary standards such as Community and Biodiversity Standards, the Gold Standard and the Climate whose aim is to attract such investors who are eager to pay a premium for emission reductions with assured co-profits (Nussbaumer, 2009). The Gold Standard is applicable to RE and energy efficiency projects. The most popular RE projects include biomass energy, wind, biogas, landfill, hydro, and solar. However, these labeled projects are only able to make up a minor share of the total volume of CDM projects (Wood, 2011). There is now enough literature addressing how to reform the CDM for better achieving novel and improved mechanisms for SD. Ideas consist of an up-scaling of mitigation arrangements with the help of sector no-lose targets (Ward, 2008), differentiation of unindustrialized country admissibility for CDM crediting (Murphy et al., 2008), introduction of novel sectoral methods and structural alterations for the CDM to play a part in long term paybacks for a low-carbon economy (Americano, 2008).

Approaches for SD can also be addressed from a broader viewpoint than sustainability assessments. A particularly attractive idea is that as compared to developed countries (DCs), under DCs might be able to follow more low-carbon and SD pathways. However, such decisions are both societal as well as political, and are intrinsically dependent on the understanding of the leapfrogging model.

4.5.3. Integrating Renewable Energy (RE) and Sustainable Development (SD) Strategies

There are two approaches through which opportunities for RE can play a part in NSDS:

- Integration of RE and sustainable development goals (SDGs) into development plans and policies like Poverty Reduction Strategy Plans and budgeting processes; and
- By developing sectoral strategies for RE playing a part in goals for low-carbon, green growth, and SD.

Despite the idea of NSDS being born at the international level, still the application of strategies in reality takes place at the national level. By 2009, a total of 106 countries matching to 55% of United Nations Member States had testified to the Commission on SD that they were carrying out NSDS. The whole notion of NSDS is the integration of principles for SD as well as targeting, resourcing, and monitoring (that is, the monitoring and measurement of development results) into a country's prevailing development process (George and Kirkpatrick, 2006). NSDS need not to be a novel, separate strategy but are supposed to integrate SD issues into a country's prevailing decision-making and governance framework. Since different countries greatly vary in their developmental, institutional, as well as geographical conditions, therefore, there is no blueprint present for NSDS, however commonly; they are structured into following three levels:

- Major/Chief Goals and policy areas for instance, dealing with energy security and climate change;
- Tangible objectives and concerns like energy efficiency, transport, and RE; and
- Aims and actions, for instance, liberalizing energy markets, implementing a RE strategy, or making use of the CDM in order to support small RE power projects.

In reality, there has been little record of progress when it comes to implementation of NSDS. Volkery et al. (2006) observed that numerous countries are still at initial phases of learning and a major challenge is harmonization of NSDS with various other strategy practices, for instance, the national budget, sub-national, and sectoral strategy procedures. In majority of countries, the NSDS only provides a summary of prevailing strategies and by itself, it works as a post-rationalization instead of an overarching framework stimulating and guiding new action. In comparison with the rich institutional landscape for economic development and cooperation, in case of SD, the institutional landscape is still small however; it can be improved by means of better ownership of SD strategies central to government.

RE strategies for green, low-carbon, and SD are becoming more significant as a way to achieve goals, for example, energy security, GHG concentration stabilization, energy access for the deprived and the establishment of green jobs (Lund et al., 2011). By 2010, worldwide more than 84 countries had taken on policy targets for the share of RE; normally 5 to 30% for producing electricity. In China, targets for final energy is 15% by 2020, in Europe, it is 20% by 2020 and was 100% by 2013 in the island states of Tonga and Fiji (REN21, 2010; SARI, 2010).

Basically, RE strategies define the challenges and potential solutions of phasing out such technologies and fossil fuels which are not sustainable; meanwhile phasing in RE systems (Lund, 2007; Verbruggen and Lauber, 2009). In order to harness the broad potential of RE sources, significant technological alterations are required together with such regulation and policies to ensure an effective, sustainable, as well as efficient usage of energy technologies and sources. We need comprehensive scientific qualification and differentiation of renewable electricity technologies and sources in order to assess the vast diversity in the field.

4.6. PRIVATE, LOCAL, AND NONGOVERNMENTAL SUSTAINABLE DEVELOPMENT (SD) INITIATIVES

At the limited level, local governments and cities in coalition with different scale businesses as well as citizen interests can prove to be the drivers of change for RE deployment (REN21, 2009). In retort to empowering framework, conditions at both national and international levels, local governments and cities can individually utilize their purchasing and legislative power for implementation of different RE initiatives in their personal operations as well as in the wider community. In general, local policy initiatives are driven through various sustainability goals like the share of renewable electricity manufacture, low GHG concentration stabilization, or total consumption of energy (Ostergaard and Lund, 2010). Other categories of local RE policies and SD initiatives are urban planning incorporating RE, annexation of RE in building codes or allowing, regulatory measures, for instance, blending of biofuels, RE application in municipal infrastructure and operations as well as voluntary movements to support RE to serve as a role model for both citizens and business (REN21, 2009, 2010). For purpose of sharing experiences as well as inspiring local actions, a wide variety of networks and initiatives have been materialized like the Local Government Climate Roadmap, World Mayors and Local Governments Climate Protection Agreement, 100% RE regions, Solar Cities, ICLEI's Local Renewables Initiative, Green Capital Awards, the European Green Cities Network, and several others. RE can also bring various SD benefits like a local supply of energy, which will save both money and energy, and create local jobs which will involve the private sector as well to play their role in providing RE services (Del Rio and Burguillo, 2008, 2009).

If community-based organizations are involved, it can greatly help to reduce the local opposition to RE installations by means of helping in local ownership and sharing of profits (Rogers et al., 2008; Zografakis et al., 2009). The formation of local energy markets can greatly help in providing opportunities for local private investors (Hvelplund, 2006) and hence can ensure public approval of integrating a growing quantity of local RE installations (solar panels, windmills, biogas plants, etc.) into the energy system. Progressive effects on the local economy can further improve public outlooks towards RE developments (Maruyama et al., 2007; Aitken, 2010; Warren and McFadyen, 2010). Different case studies when carried out in Germany and France for evaluation of the success of wind energy projects stated that the acquaintance of the developer with local situations and apprehensions as well as provision of information, transparency, and participation of the general public in the planning procedure from the initial stages on are key factors in increasing public acceptance towards RE (Wolsink, 2007). Regarding developing countries, this comprises the empowerment of rural women as well to strive for the best solutions for meeting community energy requirements (Omer, 2003; Oikonomou et al., 2009; A. Singh, 2009).

REFERENCES

1. Afgan, N. H., Begic, F., & Kazagic, A., (2007). Multi-criteria sustainability assessment: A tool for evaluation of new energy system. *Thermal Science*, *11*(3), 43–53.
2. Aitken, M., (2010). Wind power and community benefits: Challenges and opportunities. *Energy Policy*, *38*(10), 6066–6075.
3. Americano, B., (2008). CDM in Brazil: Towards structural change for sustainable development in some sectors. In: Olsen, K. H., & Fenhann, J., (eds.), *A Reformed CDM* (Vol. 1, pp. 23–46). UNEP Risø Centre, Roskilde, Denmark.
4. Angel-Urdinola, D., & Wodon, Q., (2007). Do utility subsidies reach the poor? Framework and evidence for Cape Verde, Sao Tome, and Rwanda. *Economics Bulletin*, *9*(4), 1–7.
5. Bailis, R., & Cutler, J. C., (2004). Wood in household energy use. In: *Encyclopedia of Energy* (Vol. 1, pp. 509–526). Elsevier, New York, NY, USA.
6. Bailis, R., Cowan, A., Berrueta, V., & Masera, O., (2009). Arresting the killer in the kitchen: The promises and pitfalls of commercializing improved cook stoves. *World Development*, *37*(10), 1694–1705.
7. Bailis, R., Ezzati, M., & Kammen, D. M., (2005). Mortality and greenhouse gas impacts of biomass and petroleum energy futures in Africa. *Science, 308*(5718), 98–103.
8. Bang, H. K., Ellinger, A. E., Hadjimarcou, J., & Traichal, P. A., (2000). Consumer concern, knowledge, belief, and attitude toward renewable energy: An application of the reasoned action theory. *Psychology and Marketing*, *17*(6), 449–468.
9. Bao, G. J., (2010). Study on the relevance of cultural system and hydropower resettlement project in Nujiang River. In: *Advance in Resources and Environmental Economics Research* (Vol. 1, pp. 360–364). Scientific Research Publishing, California, USA.
10. Barbier, E. B., (2009). *Rethinking the Economic Recovery: A Global Green New Deal* (Vol. 1, pp. 1–30). Report prepared for the Green Economy Initiative and the Division of Technology, Industry and Economics of the UN Environment Programme. Available at: http://www.unep.org/greeneconomy/portals/30/docs/GGND-Report-April2009.pdf (accessed on 1 May 2020).

11. Barnes, D., & Halpern, J., (2001). Reaching the poor: Designing energy subsidies to benefit those that need it. *Refocus*, *2*(6), 32–34.
12. Becerra-Lopez, H. R., & Golding, P., (2008). Multi-objective optimization for capacity expansion of regional power-generation systems: Case study of far west Texas. *Energy Conversion and Management*, *49*(6), 1433–1445.
13. Bell, D., Gray, T., & Haggett, C., (2005). The 'social gap' in wind farm siting decisions: Explanations and policy responses. *Environmental Politics*, *14*(4), 460–477.
14. Ben, H. E., (2003). Woodfuels consumption in Ghana: Environmental issues and challenges. In: *Proceedings of 1st International Conference on Energy and Environment* (pp. 495–500). Changsha, China.
15. Bishop, J. D. K., & Amaratunga, G. A. J., (2008). Evaluation of small wind turbines in distributed arrangement as sustainable wind energy option for Barbados. *Energy Conversion and Management*, *49*(6), 1652–1661.
16. Brent, A. C., & Kruger, W. J. L., (2009). Systems analyses and the sustainable transfer of renewable energy technologies: A focus on remote areas of Africa. *Renewable Energy*, *34*(7), 1774–1781.
17. Brent, A. C., & Rogers, D. E., (2010). Renewable rural electrification: Sustainability assessment of mini-hybrid off-grid technological systems in the African context. *Renewable Energy*, *35*(1), 257–265.
18. Brewer, T. L., (2004). The WTO and the Kyoto Protocol: Interaction issues. *Climate Policy*, *4*, pp. 3–12.
19. Brophy, P., (1997). Environmental advantages to the utilization of geothermal energy. *Renewable Energy*, *10*, pp. 367–377.
20. Brown, P. H., & Xu, K., (2010). Hydropower development and resettlement policy on China's Nu River. *Journal of Contemporary China*, *19*(66), 777–797.
21. Brunnschweiler, C. N., (2010). Finance for renewable energy: An empirical analysis of developing and transition economies. *Environment and Development Economics, 15*(3), 241–274.
22. Cavallaro, F., (2009). Multi-criteria decision aid to assess concentrated solar thermal technologies. *Renewable Energy*, *34*(7), 1678–1685.
23. Cernea, M., (1997). The risks and reconstruction model for resettling displaced populations. *World Development*, *25*(10), 1569–1587.

24. Chaurey, A., & Kandpal, T. C., (2010). Assessment and evaluation of PV based decentralized rural electrification: An overview. *Renewable and Sustainable Energy Reviews, 14*(8), 2266–2278.

25. Chen, S. J., (2009). Discussion of resettlement cost externalization of water resources and hydropower projects. *Advances in Water Resources and Hydraulic Engineering, 1–6*, pp. 1427–1432.

26. Coady, D., Gillingham, R., Ossowski, R., Piotrowski, J., Shamsuddin, T., & Tyson, J., (2010). *Petroleum Product Subsidies: Costly, Inequitable, and Rising* (Vol. 1, pp. 1–49). IMF Staff Position Note SPN/10/05, International Monetary Fund, Washington, DC, USA.

27. Coady, D., Grosh, M., & Hoddinott, J., (2004). *Targeting of Transfers in Developing Countries: Review of Lessons and Experience* (Vol. 1, pp. 1–39). The World Bank, Washington, DC, USA.

28. Cowan, K. R., Daim, T., Wakeland, W., Fallah, H., Sheble, G., Lutzenhiser, L., Ingle, A., Hammond, R., & Nguyen, M., (2009). Forecasting the adoption of emerging energy technologies: Managing climate change and evolving social values. In: *Portland International Conference on Management of Engineering and Technology, PICMET 2009* (Vol. 1, pp. 2964–2974). Portland, OR, USA.

29. Creutzig, F., & He, D., (2009). Climate change mitigation and co-benefits of feasible transport demand policies in Beijing. *Transportation Research Part D: Transport and Environment, 14*(2), 120–131.

30. Davidson, O., K. Halsnæs, Huq, S., Kok, M., Metz, B., Sokona, Y., & Verhagen, J., (2003). The development and climate nexus: The case of sub Saharan Africa. *Climate Policy*, 3(1), S97–S113

31. Deichmann, U., Meisner, C., Murray, S., & Wheeler, D., (2011). The economics of renewable energy expansion in rural sub-Saharan Africa. *Energy Policy, 39*(1), 215–227.

32. Del Granado, J. A., Coady, D., & Gillingham, R., (2010). *The Unequal Benefits of Fuel Subsidies: A Review of Evidence for Developing Countries* (IMF Working Paper WP/10/202, pp. 1–40) International Monetary Fund, Washington, DC, USA.

33. Del Rio, P., & Burguillo, M., (2008). Assessing the impact of renewable energy deployment on local sustainability: Towards a theoretical framework. *Renewable and Sustainable Energy Reviews, 12*(5), 1325–1344.

34. Del Rio, P., & Burguillo, M., (2009). An empirical analysis of the impact of renewable energy deployment on local sustainability. *Renewable and Sustainable Energy Reviews*, *13*(6/7), 1314–1325.
35. Devine-Wright, P., (2005). Beyond Nimbyism: Towards an integrated framework for understanding public perceptions of wind energy. *Wind Energy*, *8*(2), 125–139.
36. Devine-Wright, P., (2009). Rethinking Nimbyism: The role of place attachment and place identity in explaining place-protective action. *Journal of Community and Applied Social Psychology*, *19*(6), 426–441.
37. Donat, C. L., Gil-Gonzalez, D., Alvarez-Dardet, C. D., and Hernandez, A. I., (2010). The environmental millennium development goal: progress and barriers to its achievement. *Environmental Science and Policy*, *13*(2), 154–163.
38. Eason, T. N., Owusu, Y. A., & Chapman, H., (2009). A systematic approach to assessing the sustainability of the renewable energy standard (RES) under the proposed American renewable energy act (H.R. 890). *International Journal of Global Energy Issues*, *32*(1/2), 139–159.
39. Eberhard, A., Foster, V., Briceño-Garmendia, C., Ouedraogo, F., Camos, D., & Shkaratan, M., (2008). *Underpowered: The State of the Power Sector in Sub-Saharan Africa* (Vol. 1, pp. 1–29). World Bank, Washington, DC, USA.
40. Engfeldt, L. G., (2009). *From Stockholm to Johannesburg and Beyond* (Vol. 1, pp. 1–49). Press, Information and Communication Department, Ministry for Foreign Affairs, Stockholm, Sweden.
41. ESMAP, (2007). *Technical and Economic Assessment of Off-Grid, Mini-grid and Grid Electrification Technologies* (Vol. 1, pp.1–49). Energy Sector Management Assistance Program, World Bank, Washington, DC, USA.
42. Evans, A., Strezov, V., & Evans, T. J., (2009). Assessment of sustainability indicators for renewable energy technologies. *Renewable and Sustainable Energy Reviews*, *13*(5), 1082–1088.
43. Fahlen, E., & Ahlgren, E. O., (2010). Accounting for external costs in a study of a Swedish district-heating system: An assessment of environmental policies. *Energy Policy*, *38*(9), 4909–4920.

44. Figueiredo, P. N., (2003). Learning, capability accumulation and firms differences: Evidence from latecomer steel. *Industrial and Corporate Change, 12*(3), 607–643.
45. Figueres, C., & Streck, C., (2009). The evolution of the CDM in a post-2012 climate agreement. *Journal of Environment and Development, 18*, pp. 227–246.
46. Founex Committee, (1971). *The Founex Report on Development and Environment* (Vol. 1, pp.1–29). Founex Committee, Founex, Switzerland.
47. G-20, (2009). *Leaders' Statement: The Pittsburgh Summit: The Pittsburgh G-20 Summit* (Vol. 1, pp. 1–69). U.S. Department of State, Pittsburgh, PA, USA.
48. Gangopadhyay, S., Ramaswami, B., & Wadhwa, W., (2005). Reducing subsidies on household fuels in India: How will it affect the poor? *Energy Policy, 33*(18), 2326–2336.
49. George, C., & Kirkpatrick, C., (2006). Assessing national sustainable development strategies: Strengthening the links to operational policy. *Natural Resources Forum, 30*(2), 146–156.
50. GNESD, (2004). *Energy Access: Making Power Sector Reform Work for the Poor* (Vol. 1, pp.1–42). Global Network on Energy for Sustainable Development (GNESD), Roskilde, Denmark.
51. GNESD, (2007a). *Reaching the Millennium Development Goals and Beyond: Access to Modern Forms of Energy as a Prerequisite* (Vol. 1, pp. 1–49). Global Network on Energy for Sustainable Development (GNESD), Roskilde, Denmark.
52. GNESD, (2007b). *Renewable Energy Technologies and Poverty Alleviation: Overcoming Barriers and Unlocking Potentials* (Vol. 1, pp. 1–29). Global Network on Energy for Sustainable Development (GNESD), Roskilde, Denmark.
53. GNESD, (2008). *Clean Energy for the Urban Poor: An Urgent Issue* (Vol. 1, pp. 1–49). Global Network on Energy for Sustainable Development (GNESD), Roskilde, Denmark.
54. GNESD, (2010). *Energy Security* (Vol. 1, pp. 1–44). Global network on energy for sustainable development (GNESD), Roskilde, Denmark.
55. Goldemberg, J., & Teixeira, S. C., (2004). Renewable energy: Traditional biomass vs. modern biomass. *Energy Policy, 32*(6), 711–714.

56. Goldemberg, J., Coelho, S. T., & Guardabassi, P., (2008). The sustainability of ethanol production from sugarcane. *Energy Policy*, *36*(6), 2086–2097.

57. Green, R., & Vasilakos, N., (2011). The economics of offshore wind. *Energy Policy*, *39*(2), 496–502.

58. Gross, C., (2007). Community perspectives of wind energy in Australia: The application of a justice and community fairness framework to increase social acceptance. *Energy Policy*, *35*(5), 2727–2736.

59. Gunawardena, U. A. D. P., (2010). Inequalities and externalities of power sector: A case of Broadlands hydropower project in Sri Lanka. *Energy Policy*, *38*(2), 726–734.

60. Harmon, R. R., & Cowan, K. R., (2009). A multiple perspectives view of the market case for green energy. *Technological Forecasting and Social Change*, *76*(1), 204–213.

61. Hennenberg, K. J., Dragisic, C., Haye, S., Hewson, J., Semroc, B., Savy, C., Wiegmann, K., Fehrenbach, H., & Fritsche, U. R., (2010). The power of bioenergy-related standards to protect biodiversity. *Conservation Biology*, *24*(2), 412–423.

62. Hreinsson, E. B., (2007). Environmental, technical, economics, and policy issues of the master plan for the renewable hydro and geothermal energy resources in Iceland. In: *42nd Universities Power Engineering Conference* (Vols. 1–3, pp. 726–731), Brighton, UK.

63. Hvelplund, F., (2006). Renewable energy and the need for local energy markets. *Energy*, *31*(13), 2293–2302.

64. IEA/OECD/World Bank, (2010). *The Scope of Fossil-Fuel Subsidies in 2009 and a Roadmap for Phasing Out Fossil Fuel Subsidies* (Vol. 1, pp.1–48). International Energy Agency (IEA), Organization for Economic Co-Operation and Development (OECD), and the World Bank. Paris, France and Washington, DC, USA. Available at: www.iea.org/weo/docs/second_joint_report.pdf (accessed on 1 May 2020).

65. IIED, (2009). *'Land grabs' in Africa: Can the Deals Work for Development? September Briefing* (Vol. 1, pp. 1–55). International Institute for Environment and Development (IIED), London, UK.

66. Jager, W., (2006). Stimulating the diffusion of photovoltaic systems: A behavioral perspective. *Energy Policy*, *34*(14), 1935–1943.

67. Jobert, A., Laborgne, P., & Mimler, S., (2007). Local acceptance of wind energy: Factors of success identified in French and German case

studies. *Energy Policy*, *35*(5), 2751–2760.

68. Karekezi, S., & Kithyoma, W., (2003). Renewable energy in Africa: Prospects and limits. In: *The Republic of Senegal and United Nations Workshop for African Energy Experts on Operationalizing the NEPAD Energy Initiative* (Vols. 2–4, p. 30). Dakar, Senegal. Available at: www.un.org/esa/sustdev/sdissues/energy/op/nepadkarekezi (accessed on 1 May 2020).
69. Karekezi, S., Kimani, J., Onguru, O., & Kithyoma, W., (2009). *Large Scale Hydropower, Renewable Energy and Adaptation to Climate Change. Climate Change and Energy Security in East and Horn of Africa* (Vol. 1, pp. 1–39). Energy, Environment and Development Network for Africa, Nairobi, Kenya.
70. Kassenga, G. R., (2008). The status and constraints of solar photovoltaic energy development in Tanzania. *Energy Sources Part B-Economics Planning and Policy*, *3*(4), 420–432.
71. Kowalski, K., Stagl, S., Madlener, R., & Omann, I., (2009). Sustainable energy futures: Methodological challenges in combining scenarios and participatory multi-criteria analysis. *European Journal of Operational Research*, *197*(3), 1063–1074.
72. Kumar, P., Britter, R., & Gupta, N., (2009). Hydrogen fuel: Opportunities and barriers. *Journal of Fuel Cell Science and Technology*, *6*(2), 021009. doi: 10.1115/1.3005384.
73. Ladenburg, J., (2010). Attitudes towards offshore wind farms: The role of beach visits on attitude and demographic and attitude relations. *Energy Policy*, *38*(3), 1297–1304.
74. Lall, S., (2002). Linking FDI and technology development for capacity building and strategic competitiveness. *Transnational Corporations*, *11*(3), 39–88.
75. Limmeechokchai, B., & Chawana, S., (2007). Sustainable energy development strategies in the rural Thailand: The case of the improved cooking stove and the small biogas digester. *Renewable and Sustainable Energy Reviews*, *11*(5), 818–837.
76. Litvine, D., & Wüstenhagen, R., (2011). Helping "light green" consumers walk the talk: Results of a behavioral intervention survey in the Swiss electricity market. *Ecological Economics*, *70*(3), 462–474.
77. Lucon, O., & Rei, F., (2006). *Identifying Complementary Measures to Ensure the Maximum Realization of Benefits from the Liberalization of*

EG&S. Case study: Brazil (Vol. 1, p. 35). Working Paper No. 2004–04, COM/ENV/TD(2003)116/FINAL, Organization for Economic Co-operation and Development, Trade and Environment, Paris, France. Available at: www.oecd.org/dataoecd/18/53/37325499.pdf (accessed on 1 May 2020).

78. Lund, H., (2007). Renewable energy strategies for sustainable development. *Energy*, *32*(6), 912–919.
79. Lund, H., Ostergaard, P. A., & Stadler, I., (2011). Towards 100% renewable energy systems. *Applied Energy*, *88*(2), 419–421.
80. Mahapatra, S., Chanakya, H. N., & Dasappa, S., (2009). Evaluation of various energy devices for domestic lighting in India: Technology, economics, and CO_2 emissions. *Energy for Sustainable Development*, *13*(4), 271–279.
81. Mainali, B., & Silveira, S., (2011). Financing off-grid rural electrification: Country case Nepal. *Energy*, *36*(4), 2194–2201.
82. Malesios, C., & Arabatzis, G., (2010). Small hydropower stations in Greece: The local people's attitudes in a mountainous prefecture. *Renewable and Sustainable Energy Reviews*, *14*(9), 2492–2510.
83. Maruyama, Y., Nishikido, M., & Iida, T., (2007). The rise of community wind power in Japan: Enhanced acceptance through social innovation. *Energy Policy*, *35*(5), 2761–2769.
84. Mathews, J. A., (2008). How carbon credits could drive the emergence of renewable energies. *Energy Policy*, *36*(10), 3633–3639.
85. McGowan, F., & Sauter, R., (2005). *Public Opinion on Energy Research: A Desk Study for the Research Councils* (Vol. 1, pp. 1–44). Sussex Energy Group, Science and Technology Policy Research, University of Sussex, Brighton, UK.
86. Michalena, E., Hills, J., & Amat, J. P., (2009). Developing sustainable tourism, using a multicriteria analysis on renewable energy in the Mediterranean Islands. *Energy for Sustainable Development*, *13*(2), 129–136.
87. Mirza, U. K., Ahmad, N., Harijan, K., & Majeed, T., (2009). Identifying and addressing barriers to renewable energy development in Pakistan. *Renewable and Sustainable Energy Reviews*, *13*(4), 927–931.
88. Modi, V., McDade, S., Lallement, D., & Saghir, J., (2005). *Energy Services for the Millennium Development Goals* (Vol. 1, pp. 1–33). Energy Sector Management Assistance Programme, United Nations

Development Programme, UN Millennium Project and World Bank, New York, NY, USA.

89. Mondal, M. A. H., Kamp, L. M., & Pachova, N. I., (2010). Drivers, barriers, and strategies for implementation of renewable energy technologies in rural areas in Bangladesh: An innovation system analysis. *Energy Policy*, *38*(8), 4626–4634.

90. Monroy, C. R., & Hernandez, A. S. S., (2008). Strengthening financial innovation energy supply projects for rural communities in developing countries. *International Journal of Sustainable Development and World Ecology*, *15*(5), 471–483.

91. Moore, D., Dore, J., & Gyawali, D., (2010). The World Commission on dams + 10: Revisiting the large dam controversy. *Water Alternatives*, *3*(2), 3–13.

92. Murphy, D., Cosbey, A., & Drexhage, J., (2008). Market mechanisms for sustainable development in a post-2012 climate regime: Implications for the development dividend. In: Olsen, K. H., & Fenhann, J., (eds.), *A Reformed CDM-Including New Mechanisms for Sustainable Development* (Vol. 1, pp. 9–23). UNEP Risø Centre, Roskilde, Denmark. Available at: www.cd4cdm.org/Publications/Perspectives/ReformedCDM.pdf (accessed on 1 May 2020).

93. Nandy, S., Kushwaha, S. P. S., & Mukhopadhyay, S., (2007). Monitoring the Chilla-Motichur wildlife corridor using geospatial tools. *Journal for Nature Conservation*, *15*(4), 237–244.

94. Nannen, V., & Van den Bergh, J. C. J. M., (2010). Policy instruments for evolution of bounded rationality: Application to climate-energy problems. *Technological Forecasting and Social Change*, *77*(1), 76–93.

95. Neudoerffer, R. C., Malhotra, P., & Ramana, P. V., (2001). Participatory rural energy planning in India: A policy context. *Energy Policy*, *29*(5), 371–381.

96. Nieuwenhout, F. D. J., Van Dijk, A., Van Dijk, V. A. P., Hirsch, D., Lasschuit, P. E., Van Roekel, G., Arriaza, H., Hankins, M., Sharma, B. D., & Wade, H., (2000). *Monitoring and Evaluation of Solar Home Systems. Experiences with Applications of Solar PV for Households in Developing Countries* (Vol. ECN-C--00–089, pp. 1–60). Report, Netherlands Energy Research Foundation, Department of Science, Technology and Society of Utrecht University, Utrecht, The Netherlands.

97. OECD, (2002). *Governance for Sustainable Development: Five OECD Case Studies*. Organization for Economic Co-operation and Development (OECD), Paris, France.
98. OECD, (2006). *Environmental and Energy Products: The Benefits of Liberalizing Trade* (Vol. 1, pp. 1–55). Organization for Economic Co-operation and Development (OECD), Paris, France (ISBN-92–64–02481–6).
99. Oikonomou, E. K., Kilias, V., Goumas, A., Rigopoulos, A., Karakatsani, E., Damasiotis, M., Papastefanakis, D., & Marini, N., (2009). Renewable energy sources (RES) projects and their barriers on a regional scale: The case study of wind parks in the Dodecanese islands, Greece. *Energy Policy*, *37*(11), 4874–4883.
100. Olsen, K. H., & Fenhann, J., (2008a). *A Reformed CDM Including New Mechanisms for Sustainable Development* (Vol. 1, p. 184). UNEP Risø Centre, Roskilde, Denmark.
101. Ölz, S., & Beerepoot, M., (2010). *Deploying Renewables in Southeast Asia: Trends and Potentials* (Vol. 1, pp. 1–49). Organization for Economic Co-operation and Development and International Energy Agency, Paris, France.
102. Omer, A. M., (2003). Implications of renewable energy for women in Sudan: Challenges and opportunities. *International Journal of Sustainable Development*, *6*(2), 246–259.
103. Onat, N., & Bayar, H., (2010). The sustainability indicators of power production systems. *Renewable and Sustainable Energy Reviews*, *14*(9), 3108–3115.
104. Ostergaard, P. A., & Lund, H., (2010). A renewable energy system in Frederikshavn using low-temperature geothermal energy for district heating. *Applied Energy*, *88*(2), 479–487.
105. Owen, A. D., (2006). Renewable energy: Externality costs as market barriers. *Energy Policy*, *34*(5), 632–642.
106. Painuly, J. P., (2001). Barriers to renewable energy penetration, a framework for analysis. *Renewable Energy*, *24*(1), 73–89.
107. Paulsson, E., (2009). A review of the CDM literature: From fine-tuning to critical scrutiny? *International Environmental Agreements-Politics Law and Economics*, *9*(1), 63–80.
108. Petersen, L. K., & Andersen, A. H., (2009). *Socio-cultural Barriers to the Development of a Sustainable Energy System: The Case of Hydrogen*

(Vol. 1, pp. 1–49). National Environmental Research Institute, Aarhus, Denmark.

109. Pohekar, S. D., & Ramachandran, M., (2006). Multi-criteria evaluation of cooking devices with special reference to utility of parabolic solar cooker (PSC) in India. *Energy*, *31*(8/9), 1215–1227.

110. Rao, K. U., & Kishore, V. V. N., (2010). A review of technology diffusion models with special reference to renewable energy technologies. *Renewable and Sustainable Energy Reviews*, *14*(3), 1070–1078.

111. Ravindranath, N. H., & Balachandra, R., (2009). Sustainable bioenergy for India: Technical, economic and policy analysis. *Energy*, *34*(8), 1003–1013.

112. Reddy, S., & Painuly, J. P., (2004). Diffusion of renewable energy technologies-barriers and stakeholders' perspectives. *Renewable Energy*, *29*(9), 1431–1447.

113. REN21, (2009). *Renewables Global Status Report: 2009 Update* (Vol. 1, pp. 39–42). Renewable Energy Policy Network for the 21st Century Secretariat, Paris, France.

114. REN21, (2010). *Renewables 2010: Global Status Report* (Vol. 1, pp. 76–80). Renewable Energy Policy Network for the 21st Century Secretariat, Paris, France.

115. Richter, B. D., Postel, S., Revenga, C., Scudder, T., Lehner, B., Churchill, A., & Chow, M., (2010). Lost in development's shadow: The downstream human consequences of dams. *Water Alternatives*, *3*(2), 14–42.

116. Rogers, J. C., Simmons, E. A., Convery, I., & Weatherall, A., (2008). Public perceptions of opportunities for community-based renewable energy projects. *Energy Policy*, *36*(11), 4217–4226.

117. Rylands, A. B., & Brandon, K., (2005). Brazilian protected area. *Conservation Biology*, *19*(3), 612–618.

118. SARI, (2010). *Unlocking South Africa's Green Growth Potential: The South African Renewables Initiative* (Vol. 1, pp. 1–49). South African Renewables Initiative (SARI), Department of Trade and Industry, Department for Public Enterprises, Pretoria, South Africa.

119. Singh, A., (2009). The sustainable development of Fiji's energy infrastructure: A status report. *Pacific Economic Bulletin*, *24*(2), 141–154.

120. Sovacool, B. K., & Hirsh, R. F., (2009). Beyond batteries: An examination of the benefits and barriers to plug-in hybrid electric vehicles (PHEVs) and a vehicle-to-grid (V2G) transition. *Energy Policy*, *37*(3), 1095–1103.

121. Sovacool, B. K., (2009). The cultural barriers to renewable energy and energy efficiency in the United States. *Technology in Society*, *31*(4), 365–373.

122. Steenblik, R., (2005). *Liberalization of Trade in Renewable-Energy Products and Associated Goods: Charcoal, Solar Photovoltaic Systems and Wind Pumps and Turbines* (Vol. 1, pp. 1–59). COM/ENV/TD(2005)23/FINAL, Organization for Economic Co-operation and Development, Paris, France.

123. Steurer, R., & Martinuzzi, A., (2007). From environmental plans to sustainable development strategies. *European Environment*, *17*(3), 147–151.

124. Sutter, C., & Parreño, J. C., (2007). Does the current clean development mechanism (CDM) deliver its sustainable development claim? An analysis of officially registered CDM projects. *Climatic Change*, *84*(1), 75–90.

125. Tamiotti, L., Olhoff, A., Teh, R., Simmons, B., Kulaçoğlu, V., & Abaza, H., (2009). *Trade and Climate Change* (Vol. 1, pp. 1–49). A report by the United Nations environment programme and the World Trade Organization, World Trade Organization, Geneva, Switzerland.

126. Thiam, D. R., (2010). Renewable decentralized in developing countries: Appraisal from microgrids project in Senegal. *Renewable Energy*, *35*(8), 1615–1623.

127. Thompson, S., & Duggirala, B., (2009). The feasibility of renewable energies at an off-grid community in Canada. *Renewable and Sustainable Energy Reviews*, *13*(9), 2740–2745.

128. UN, (2002). *Report of the World Summit on Sustainable Development* (Vol. 1, pp. 170–173). A/CONF.199/20*, United Nations, Johannesburg, South Africa and New York, NY, USA.

129. UNCED, (1992). *Agenda 21* (Vol. 1, pp. 1–49). UN Conference on Environment and Development (UNCED), UN Department of Economic and Social Affairs, New York, NY, USA.

130. UNDESA, (2008). *Addressing Climate Change in National Sustainable Development Strategies-Common Practices* (Vol. 12, pp.

1–65). Background, DESA/DSD/2008/12, Commission on Sustainable Development, UN Department of Economic and Social Affairs (UNDESA), New York, NY, USA

131. Unruh, (2000). Understanding carbon lock-in. *Energy Policy*, *28*, pp. 817–830.
132. Upreti, B. R., (2004). Conflict over biomass energy development in the United Kingdom: Some observations and lessons from England and Wales. *Energy Policy*, *32*(6), 785–800.
133. Van Alphen, K., Van Sark, W. G. J. H. M., & Hekkert, M. P., (2007). Renewable energy technologies in the Maldives-determining the potential. *Renewable and Sustainable Energy Reviews*, *11*(8), 1650–1674.
134. Varun, R. P., & Bhat, I. K., (2010). A figure of merit for evaluating sustainability of renewable energy systems. *Renewable and Sustainable Energy Reviews*, *14*(6), 1640–1643.
135. Verbruggen, A., & Lauber, V., (2009). Basic concepts for designing renewable electricity support aiming at a full-scale transition by 2050. *Energy Policy*, *37*(12), 5732–5743.
136. Volkery, A., Swanson, D., Jacob, K., Bregha, F., & Pintér, L., (2006). Coordination, challenges, and innovations in 19 national sustainable development strategies. *World Development*, *34*(12), 2047–2063.
137. Walker, G., Devine-Wright, P., Hunter, S., High, H., & Evans, B., (2010). Trust and community: Exploring the meanings, contexts, and dynamics of community renewable energy. *Energy Policy*, *38*(6), 2655–2663.
138. Ward, M., (2008). Sector no-lose targets: A new scaling up mechanism for developing countries. In: Olsen, K. H., & Fenhann, J., (eds.), *A Reformed CDM Including New Mechanisms for Sustainable Development* (Vol. 1, pp. 147–163). UNEP Risø Centre, Roskilde, Denmark.
139. Warren, C. R., & McFadyen, M., (2010). Does community ownership affect public attitudes to wind energy? A case study from south-west Scotland. *Land Use Policy*, *27*(2), 204–213.
140. Werner, M., & Schaefer, A. I., (2007). Social aspects of a solar-powered desalination unit for remote Australian communities. *Desalination*, *203*(1–3), 375–393.

141. West, J., Bailey, I., & Winter, M., (2010). Renewable energy policy and public perceptions of renewable energy: A cultural theory approach. *Energy Policy*, *38*(10), 5739–5748.

142. Wilkins, G., (2002). *Technology Transfer for Renewable Energy: Overcoming Barriers in Developing Countries* (Vol. 1, pp. 1–49). The Royal Institute of International Affairs, Earthscan Publications, London, UK.

143. Winkler, H., Davidson, O., & Mwakasonda, S., (2005). Developing institutions for the clean development mechanism (CDM): African perspectives. *Climate Policy*, *5*, pp. 209–220.

144. Wolsink, M., (2000). Wind power and the NIMBY-myth: Institutional capacity and the limited significance of public support. *Renewable Energy*, *21*(1), 49–64.

145. Wolsink, M., (2007a). Planning of renewables schemes: Deliberative and fair decision-making on landscape issues instead of reproachful accusations of noncooperation. *Energy Policy*, *35*(5), 2692–2704.

146. Wolsink, M., (2007b). Wind power implementation: The nature of public attitudes: Equity and fairness instead of 'backyard motives.' *Renewable and Sustainable Energy Reviews*, *11*(6), 1188–1207.

147. Wolsink, M., (2010). Near-shore wind power-protected seascapes, environmentalists' attitudes, and the technocratic planning perspective. *Land Use Policy*, *27*(2), 195–203.

148. Wood, R. G., (2011). *Carbon Finance and Pro-Poor Co-benefits: The Gold Standard and Climate, Community and Biodiversity Standards* (Vol. 1, pp. 20–24). International Institute for Environment and Development, London, UK.

149. World Bank, (2001). *The World Bank Group's Energy Program-Poverty Reduction, Sustainability and Selectivity* (Vol. 1, pp. 1–59). Energy and Mining Sector Board, The World Bank, Washington, DC, USA.

150. World Bank, (2010). *World Development Indicators 2010* (Vol. 1, p. 489). World Bank, Washington, DC, USA.

151. World Bank, (2011). *Climate Impacts on Energy Systems: Key Issues for Energy Sector Adaptation: A World Bank Study* (Vol. 1, pp. 1–60). The World Bank, Washington, DC, USA (ISBN: 978–0–8213–8697–2).

152. World Commission on Dams, (2000). *Dams and Development: A New Framework for Decision-Making* (Vol. 1, pp. 1–39). Earthscan Publications Ltd, London, UK and Sterling, VA, USA.
153. WTO, (2010). *Background Note: Trade and Environment in the WTO* (Vol. 1, pp. 1–44). World Trade Organization, Geneva, Switzerland.
154. Wustenhagen, R., Wolsink, M., & Burer, M. J., (2007). Social acceptance of renewable energy innovation: An introduction to the concept. *Energy Policy*, *35*(5), 2683–2691.
155. Zoellner, J., Schweizer-Ries, P., & Wemheuer, C., (2008). Public acceptance of renewable energies: Results from case studies in Germany. *Energy Policy*, *36*(11), 4136–4141.
156. Zografakis, N., Sifaki, E., Pagalou, M., Nikitaki, G., Psarakis, V., & Tsagarakis, K. P., (2009). Assessment of public acceptance and willingness to pay for renewable energy sources in Crete. *Renewable and Sustainable Energy Reviews*, *14*(3), 1088–1095.
157. Zuk, M., Rojas, L., Blanco, S., Serrano, P., Cruz, J., Angeles, F., Tzintzun, G., Armendariz, C., Edwards, R. D., Johnson, M., Riojas-Rodriguez, H., & Masera, O., (2007). The impact of improved wood-burning stoves on fine particulate matter concentrations in rural Mexican homes. *Journal of Exposure Science and Environmental Epidemiology*, *17*(3), 224–232.

CHAPTER 5

Economic Features of Sustainable Development

CONTENTS

5.1. Introduction 144

5.2. Sustainable Growth 144

5.3. Environmental Economy 147

5.4. Socio-Economic Aspect of Sustainable Development (SD) 149

References 153

5.1. INTRODUCTION

The question related to the purpose of continued growth has been asked for more than three decades due to problems originating from environmental contemplation and the uneven division of wealth in the world. The idea of "Sustainable Development (SD)" first came into light about 25 years ago (IUCN, 1980), which was generally assumed to integrate ecological, social, and economic dynamics. Present-day, this idea is the cause of highly deviating explanation during the economic analysis. To have a better idea of SD, we first consider these deviating aspects, while keeping in mind the obstacles rise in achieving that goal. A lot of work has been done on that topic; several definitions have been proposed and aggregations and oppositions on them also debated. Many struggles have been carried out in this type of discussion (Godard, 1994; Zaccaï, 2002). Notwithstanding a few features from past research cannot be excluded, the proposed outline will be different from those discussed in other literature. Particular attention will be given on that problem relating to SD keep related to the idea of "sustainability" as the idea of "development." So, therefore, it is not a shocking thing to find writers who passionately object this concept in our studies. The debate of this unit was divided into three parts of work to give an outline of the subject. In the first work, the current economic theory approach is discussed in which sustainable growth is presented as a sufficient and necessary constraint for SD. In the second part, socio environmental conditions are tried to make, in those conditions, economic development must do occur. That part is related to the ecologist's idea of bound to growth. In the third part, the Third World briefed issues are addressed; attention is given to question related to the idea of development and social inequalities. It is important to ask a question related to the concerns of all the above views regarding the policies and objectives of sustainability, along with that answers to these concerns are required (Baumol and Oates, 1988).

5.2. SUSTAINABLE GROWTH

The countries of moderate economics are considered in the first group. Despite the fact of the idea of SD appeared, in a region, from analytics of growth-in the neo-classical corpus the main issue is that-however, the analytics who are in favor of these ideas, are trying to find a solution of these concerns. In the neo-classical thesis's response relating to these issues of SD, Solow's model with little alteration plays an important role. This acknowledged system is completed by other work and attention is given to the concept that sustainable growth is an integral part of environmental protection and environment development.

5.2.1. Growth: Necessary and Sufficient

Neoclassical analysts believe that the SD goal should consider the requirement of societies to manage the ability to contribute economic betterment with time and try to make sure that the upcoming human society also has access to the same betterment as previous societies. In this context, sustainability is defined in other words as well being of a common person does not decline with passing time. Well-being can be measured in different ways, depending on the analysis, by the quantity of common person income, consumption, and utility.

According to neoclassical experts that to achieve this objective, it is mandatory, to keep the high savings rates, so that the availability of capital stock to the next generation will be kept stable. Consequently, wealth will be allowed to flow contently over time. Economy production ability is fabricated by its stocks of usual level of training, education, skills, knowledge, and amenities, along with natural sources reserves. Harold Hotelling (1931) started a new tradition, according to which nature is a special type of capital. Despite the fact, it was just mentioned in the 1970's in the literature (Stiglitz, 1970), but at the end of the twenty century, the idea of "natural capital" started to be clearly defined and now it has constituted a part of neo-classical experts thought process for a matter of economic growth (Costanza et al., 1997). Nevertheless, the theories adopted by these experts take care the sustainability between two types of capital: to maintain the well-being and to make sure production extent of individual, an escalation in the level of "societies generated capital" should be compensated in terms of decline in the quantity of "natural capital." Therefore, a transformation occurs with time according to Robert Solow (1992:265): the current generation uses "natural capital" but in return, yields more capacity in the alternatives of skills, knowledge, and amenities and stock to upcoming generations.

To support this scheme, many theories are required. The first thing is a technical transformation which gives a pair of "contingency techniques" that cause the exchange between different types of capital. The second concern is related to the sense of a special investment regime: in 1977, Hartwick sets forth that money produced from the usage of non-renewable natural sources must be invested again in the technical capital by a specific investment fund or taxation system. The third point is that in Solow's model, prices are not presented-it promotes economy in which only one agent allocating the resources, is named as a planned economy-another theory presents by neoclassical' is that "market" should allocate the resources (Hartwich, 1971). The price system must find the value of the distinct types of capital, like the value of substitution that between them will establish.

5.2.2. Sustainable Development (SD): The Sixth Phase of Economic Growth

The neoclassical experts' argument of encouraging the previously mentioned concerns is that the environment should be protected with the concern of increasing growth rate. During the early 1970s, Wilfred Beckerman (1972) in his incisive judgment of the Meadow report put forward the most positive viewpoints related to the ability of modern economics to appropriately manage environmental issues.

In several American states, the decline in SO_2 contamination was specifically mentioned which happened even continued growth was undergoing in the country. After twenty years, the motion is placed and the campaign is reached to developing countries: Beckerman (1992) stated that "[...] there is clear evidence, although economic growth usually leads to environmental deterioration in the early stages of the process, in the end, the best—and probably the only—way to attain a decent environment in most countries is to become rich." Nevertheless, Alan Krueger and Gene Grossman (1993, 1995) who wanted to give verified evidence published an article that gave another angle to the debate on this kind of argument. These publishers wanted to develop an extensive association between contemporary environmental changes and economic growth by determining a correlation between average income (determined by using the total GDP of the countries) and actions relating to many water and air pollutants. A. Krueger and G. Grossman (1995) demonstrate a connection between economic development and economic growth in their article. Highlighting their work is the idea created by Walt Rostow (1960), according to him at some point in the human societies' history, economic development performs a role (Rostow, 1960). After that, they relish a self-sustaining growth, which turns out to be a "standard act of the economy." The network of the economy then alters with the technological advancement and its dispersion in different zones of activity; new investment opening with capital is generated by industries transformation. The inventive style of Krueger's and Grossman's idea endures in the matter that, opposite to Rostow, who had disbeliefs relating future possibility of the progressive community of his time, they exhibit them as expanding in a way that is more regardful of the environment. If Rostow is rephrased, then SD could be recognized as the "sixth stage of growth." Deserving to be remembered is that to 1960, Rostow did not restrict his views on the subject. He mentioned in his book at the last of the 1970s that he fully understood the effects of environmental issues and the reduction of natural resources on the stock ability of modern economics (Rostow, 1981). Nevertheless, for him, the development of these issues does not forecast the decline of industrial civilization. He stated that "These issues can be controlled if we put our minds to it and the solutions, we come up with will provide the foundation for the rekindling of sustainable growth." Consequently, if transformative trends for some air contaminators in the US (like SO_2 and other) are analyzed from 1970 to 1974, a reduction was observed. Conclusion of Rostow (1978:123) was that "Clearly, the national trend has been in a wholesome direction." Krueger and Grossman (1995) also endorsed the same idea; in the crux, the presented point of view is that several environmental problems solved themselves by the achievement of economic growth.

Krueger and Grossman (1995) admit that the issue is that the connection explained by the "inverted U" cannot be implemented widely. According to Kenneth Arrow et al. (1995), it is possible only for limited contaminators that have a local and short-term effect. Like, for the generation of the household leftover and release of CO_2, the outcomes are changed. The quantities increase with per capita income.

5.3. ENVIRONMENTAL ECONOMY

By examining the history of economic notion, it is observed that on the edge of the prevailing debate, there has been a precedent of depicting the ecological catastrophe that builds due to collective dynamics (Martinez-Alier, 1987; Vivien, 1994). It is also an eagerness to involve the particularity of environmental events, which cannot be lessened to the market reason that regulate this research viewpoint, which—to nominate some of its tendency, other than distinct terms picked (ecological economics, bio-economics, etc.)—could be called to mean an economy for the environment(Costanza et al., 1997). This term depicts the repeated and constant attempt made to enact economy to natural sciences since the 19th century, which is, to the establishment of a field that specially integrates economic and environmental knowledge. For the past 20 years, this struggle has mainly matured into a tendency that has been termed as "ecological economics". This term will be explained in this paper; however, this simple concept can provide varying political purposes, depending on the intended position and role given to private players or the public.

5.3.1. Prescriptive Constraint Management and Critical Natural Capital

The major factor of the economy of renewable natural resources has always been sustainability due to the reason for resource drainage. The administration of the fisheries industry and forestry industry design which were formulated at the start of the 18th century is considered as the source of the idea of SD. The forestry and fisheries, both swift developments after the 1960s. The organic resources of such industries are treated as a form of "natural capital," that's why it is essential to manage effectively over a long time (Commoner, 1971; Bourg and Erkman, 2003). The purpose to be achieved by using these bioeconomic strategies is a "maximum sustainable output," that is the maximum utilization of resources. It can be attained infinitely from the possible supply of resources. The issue is economy rationality that objects for peak profit may direct to the reduction of resources and maybe contradicting to environmental logic. This argument was put forward recently by Colin Clark (1973) with respect to fisheries economy or discussed long ago by Augustin Cournot with concern to forest management (Clark, 1973). campaigning for specific management rules and government intervention.

For the past 20 years, because of enhanced consciousness of international environmental problems, a debate on natural resources has been reaching a new direction. Nevertheless, until now the practical matter cannot hand over to an "international bio-economy" on the basis of acquired knowledge. At least, can we direct acknowledged-upon rules within the structure of ecological economics? The basic preventive rules according to Herman Daly (1990) are: (1) the renewable natural resources developing amount should be equal to amount of its regeneration; (2) the discharge amount of waste should equivalent consumption-ability of the surroundings in which it is settled; (3) Non-renewable natural resources must be utilized at the same amount at which these resources are being substituted with renewable sources. In contrast to neoclassical economists, this point of view exhibits the idea of complementarity from natural resources to other production aspects. From where the "firm sustainability" plan, which stands on the requirement to regulate overtime "critical natural capital" (Faucheux, O'Connor, 1999) stock which is essential for upcoming generations.

Despite the fact of this rule is easier said than done. Its practicalities are challenging. The first problem is to determine and judge all the extensive heterogeneous elements. The second challenge is the implantation of "prescriptive constraint management"—to adopt René Passet's (1979) phrase—on these elements, to find the constraints to the using of natural resources. Then, to explain the limits that will help the division of this restriction fairly within the system. Then, it will be important to define the organization that will grant economic experts to practice the best choice regarded to the varying limits (Schumacher, 1973; Vatin, 1998).

5.3.2. The Ecological Modernization of Capitalism and Industrial Ecology

There is another way of reasoning—industrial ecology— that motivated by environmental science concepts and theories, so that reconsider production and consumption processes. The innovation of this dialogue does not rely much on the subject of its opinion—many hypotheses described have been recognized for a long time—as it resides in those who send it. Nicholas Gallopoulos and Robert Frosch (1989), who is the writer of the article on industrial ecology, associate to the field of industry and, more accurately, to the field of engineering. Their work was mentioned in a distinct edition of Scientific American with the title of "Managing Planet Earth." This article was published before the UN Conference on Development and Environment which was held in 1992 in Rio de Janeiro and broadcast after the Brundtland Report (WCED, 1987) publication. There is nothing concurrent about this; the idea of industrial ecology appeared in the circumstances of the considerations within global environmental institutions that have orchestrated an essential part in the launch and dispersion of the idea of SD (Ronald Coase, 1960; Damian and Graz, 2001).

5.4. SOCIO-ECONOMIC ASPECT OF SUSTAINABLE DEVELOPMENT (SD)

The social problems elevated due to SD are highlighted by a third economic body. Get away from the prominent aspect, which has to deal with the coming development as a usual unfolding of social and economic history, the writers of these studies question the distinctiveness of non-development observed by some countries and opportunities of "more development" than the one following the route set by the West. Even a few desires to continue the development aims, others wish to break away and to construct other aspects of social advancement. Therefore, we are convinced to think about the leading economic morals of our wealthy rich and wealthy societies. Analyzing the concept of need, which was the point of concern and plenty of work has been written in the 1930s—and a topic emerged in Keynes's (1930) work—is coming back into the light, because of that the necessity to control global division of wealth and environmental issues.

5.4.1. Eco-Development

The managers of the Stockholm Conference (1972) introduced the phrase "eco-development"—After that, it had been seen to a foremost rift between countries of the South and countries of the North—therefore the foremost in their area of concern, would add the environment with their development aims. The eco-development strategy foreshadows as well as in parallel with the phrase "SD," which emerged at the start of the 1980s. This doctrine is developed by an economist Ignacy Sachs (1980) whose name is combined with it. The purpose of that doctrine is to tackle the distinct dynamics of rural areas economies of developing countries. But later it emerged as a generalized philosophy of development for them. The author's point of view is not to degrade the growth, but his opinion is it should help natural environments and the adequate management of resources and social progress. Sachs (1993, 2003) tried in this manner a balance, not to reach any extreme plans proposed by those who applaud nature's unlimited resources and the Malthusians. Obeying by the aspects set forward by the endogenous development theories, Sachs concludes that it is essential for every country to specify its own "development way," especially by using the "feasible technology" suitable with its ecological context, institutional, and cultural (Nordhaus, 1973; O'Rourke et al., 1996).

5.4.2. "Environmentalism of the Poor" and Unequal Exchange

For a very long time, the development economic experts explain non-development and development as the two aspects of capitalism, which flourish by making a connection of reliance between a "Periphery" and a "Centre." The South countries' economies are extroverted, connected as they are to deal done by Northern governments and multinationals, to fluctuation in global market prices and to exterior branches (Deléage, 2002; Fischer-

Kowalski, 2003). It is evident that several South countries get poorer by selling resources at cheap rates to the North countries without considering the environmental and social prices added by this type of resource production. This idea is come into light by Joan Martinez-Alier (2002), grows on the hypothesis of "uneven ecological exchange" presented by S. Amin (1976) and A. Emmanuel (1969). To prevent this problem, Herman Daly (1992)—who advise review Keynes' literature on the demand for country self-reliance—offers quit from the field of international business, his concept is confining within the eco-development ideology (Geddes, 1884; Harbaugh et al., 2002).

On a larger level, the content of J. Martinez-Alier asserts the divided conflicts connected to environmental issues. One more time, he puts poverty at the center of sustainability problems—a subject which remarkably came back during the session of Johannesburg Summit (2002)—even the situation is very different. On one side, Martinez-Alierintention's intentions are to prove that poverty is not dangerous, as hinted by Brundtland report (Joan Martinez-Alier (2002), to the environment by giving the Third World countries examples like the "environmental justice" campaign in the US or the movement started by Chico Mendès. Another side, he wants to end a misconception that environmental prevention is not just "a duty of rich." If this theory rephrased, it means the theory of "environmentalism is both for rich and poor, and latter work for a larger acceptance of their rights" (Joan Martinez-Alier, 2002).

5.4.3. De-Growth for the Future?

To make the debate more intense, some experts are thinking to not endorse the concept of development completely. As development is charged with the offense that it is being used to convert social relation into a commodity and westernization of the world. Rather than that, they are thinking to invent another plan for social shift. To achieve that objective, it will be mandatory to get benefit from Serge Latouche's (2003) expression, to build an aim of "convivial degrowth."

The idea of "degrowth" is connected to the literature of Nicholas Georgescu-Roegen (1971, 1995) and philosopher Jacques Grinevald (1974) is responsible for its distribution and reading. The energy and material constraints by the thermodynamics' laws, provoked N. Georgescu-Roegen to put forward a "minimal bioeconomic plan" which purpose is to make available material and energy supply as long as feasible for humans (Ivan

Illich, 1973). The notion on which it is proposed is that instead to act on the stock, it is appropriate to move on the need for service and good while surviving awareness of the need for poor people to improve their material status (André Gorz, 1975, 1991; Harribey, 1997).

REFERENCES

1. Amin S., (1976). *Unequal Development*. New York, Monthly Review Press.
2. Arrow, K., Bolin, B., Costanza, R., Dasgupta, P., Folke, C., Holling, C. S., Jansson, B. O., Levin, S., Mäler, K. G., Perrings, C., & Pimentel, D., (1995). Economic growth, carrying capacity, and the environment. *Science, 268*, 520–521.
3. Baumol, W. J., & Oates, W. E., (1988). *The Theory of Environmental Policy* (2nd edn., Vol. 2, pp. 11–22). Cambridge, Cambridge University Press.
4. Beckerman, W., (1972). *Economists, Scientists, and Environmental Catastrophe* (Vol. 3, pp. 327–344). Oxford Economic Papers 24.
5. Beckerman, W., (1992). *Economic Growth and the Environment: Whose Growth? Whose Environment?* (Vols. 4/20, pp. 481–496) World Development.
6. Bourg, D., & Erkman, S., (2003). *Perspectives on Industrial Ecology* (Vol. 1, pp. 1–20). Sheffield, Greenleaf Publishing.
7. Clark, C. W., (1973). Profit maximization and the extinction of animal species. *Journal of Political Economy, 81*(4), 950–961.
8. Coase, R. H., (1960). The problem of social cost. *Journal of Law and Economics, 3*, 1–44.
9. Commoner, B., (1971). *The Closing Circle: Nature, Man and Technology* (Vol. 1, pp. 2–19). New York, Knopf.
10. Costanza, R., Perrings, C., & Cleveland, C. J., (1997). *The Development of Ecological Economics*. Cheltenham, Edward Elgar.
11. Cournot, A. A., (1861). *Treatise on the Chain of Fundamental Ideas in Science and History* (Vol. 1, pp. 10–20), Paris, reprinted, Librairie J. Vrin.
12. Daly, H. E., (1990). Toward some operational principles of sustainable development. *Ecological Economics, 2*, 1–6.
13. Daly, H. E., (1992). *Free Trade, Sustainable Development and Growth: Some Serious Contradictions* (pp. 123–127). Reprinted in ecological economics and the ecology of economics, Cheltenham, Edward Elgar-1999.
14. Damian, M., & Graz, J. C., (2001). *International Trade and Sustainable Development* (Vol. 1, pp. 1–20), Paris, Economica.

15. Deléage, J. P., (2002). Scientific ecology: From nature to industry. *Ecology and Politics, 25*, 57–66.
16. Emmanuel, A., (1969). *Unequal Exchange: A Study of the Imperialism of Trade* (Vol. 1, pp. 1–18). English Trans-1972, New York & London, Monthly Review Press.
17. Faucheux, S., & O'Connor, M., (1999). *A Controversial Concept: Natural Capital* (Vol. 99–01, pp. 1–30). Cahier du C3ED, University of Versailles.
18. Fischer-Kowalski, M., (2003). On the history of industrial metabolism. In: Bourg, D., & Erkman, S., (eds.), *Perspectives on Industrial Ecology* (Vol. 1, pp. 35–45). Sheffield, Greenleaf Publishing.
19. Frosch, R., & Gallopoulos, N., (1989). Strategies for manufacturing. *Scientific American, 261*, 144–152.
20. Frosch, R., (1995). The industrial ecology of the 21st century. *Scientific American, 273*, 178–181.
21. Geddes, P., (1884). *An Analysis of the Principles of Economics* (Vol. 1, pp. 5–15). Reprinted, London, Williams and Norgate-1885.
22. Georgescu-Roegen, N., (1971). *The Entropy Law and the Economic Process* (Vol. 1, pp. 5–19). Cambridge, Harvard University Press.
23. Georgescu-Roegen, N., (1995). *The Decay* (Vol. 1, pp. 5–17). Paris, Blood of the Earth.
24. Godard, O., (1994). Sustainable development: Intellectual landscape, *Natures Sciences Sociétés, 2*(4), 309–322.
25. Gorz, A., (1975). *Ecology and Politics* (Vol. 1, pp. 5–17). Paris, Le Seuil-1978.
26. Gorz, A., (1991). *Capitalism, Socialism, Ecology* (Vol. 1, pp. 5–17). Paris, Galileo.
27. Grinevald, J., (1974). *The Economy of Decay* (Vol. 3, No. 2, pp. 69–70). Reprinted in The environmentalist.
28. Grossman, G. M., & Krueger, A. B., (1993). Environmental impacts of a north American free trade agreement. In: Garber, P. M., (ed.), *The Mexico-U.S. Free Trade Agreement* (Vol. 1, pp. 13–56). Cambridge (Mass.). The MIT Press.
29. Grossman, G. M., & Krueger, A. B., (1995). Economic growth and the environment. *Quarterly Journal of Economics, 2*, 353–377.

30. Harbaugh, W. T., Levinson, A., & Wilson, D. M., (2002). Re-examining the empirical evidence for an environmental Kuznets curve. *The Review of Economics and Statistics, 84*(3), 541–551.
31. Harribey, J. M., (1997). The economical economy. *Sustainable Development by Reducing Working Hours* (Vol. 1, pp. 5–17). Paris, L'Harmattan.
32. Hartwick, J. M., (1977). Intergenerational equity and the investing of rents from exhaustible resources. *American Economic Review, 77*(5), 972–974.
33. Hotelling, H., (1931). The economics of exhaustible economics. *Journal of Political Economy, 39*(2), 137–175.
34. Illich, I., (1973). *Tools for Conviviality* (Vol. 1, pp. 5–17). New York, Harper & Row.
35. IUCN, (1980). *World Conservation Strategy: Living Resource Conservation for Sustainable Development* (Vol. 1, pp. 5–17). UICN/PNUE/WWF, Gland.
36. Keynes, J. M., (1930). *Economic Possibilities for our Grandchildren, in Essays in Persuasion* (pp. 358–3744). New York, Harcourt Brace-1932.
37. Latouche, S., (2003). Down with sustainable development! Long live friendly degrowth!, In: Bernard, M., et al., (eds.), *(under the Coord.), Decrease Objective* (Vol. 1, pp. 19–26). Paris, Parangon.
38. Martinez-Alier, J., & Schlüpmann, K., (1987). *Ecological Economics* (Vol. 1, pp. 5–27). Oxford, Basil Blackwell.
39. Martinez-Alier, J., (2002). *The Environmentalism of the Poor* (Vol. 1, pp. 3–17). Cheltenham, Edward Elgar.
40. Meadows, D. H., et al., (1972). *The Limits to Growth* (Vol. 1, pp. 5–17). New York, Universe Books.
41. Nordhaus, W. D., (1973). *The Allocation of Energy Resources* (Vol. 3, pp. 529–576). Brooking papers on economic activity.
42. O'Rourke, D., Connelly, L., & Koshland, C., (1996). Industrial ecology: A critical review. *International Journal of Environment and Pollution, 6*, 2/3, 89–112.
43. Passet, R., (1979). *L'économique et le Vivant* (Vol. 1, pp. 1–13). Paris, Payot.

44. Passet, R., (1989). *Let the Economy Serve the Biosphere* (Vol. 1, pp. 4, 5). Le Monde diplomatique.
45. Rostow, W., (1960). *The Stages of Economic Growth: A Non-Communist Manifesto* (Vol. 1, pp. 1–19). Cambridge: Cambridge University Press.
46. Rostow, W., (1978). *Getting from Here to There* (Vol. 1, pp. 5–17). New York: McGraw-Hill.
47. Rostow, W., (1981). *Preface to the French Edition of Getting from Here to There: The Ultimatum of the Year 2000* (Vol. 1, pp. 2–17). Chances of Survival of the World Economy, trad. fse, Paris, Economica.
48. Sachs, I., (1980). *Ecodevelopment Strategies* (Vol. 1, pp. 1–17). Paris, Ed. Economics and Humanities/Ed. Ouvrières.
49. Sachs, I., (1993). *Ecodevelopment* (Vol. 1, pp. 2–19). Paris, Syros.
50. Sachs, I., (1994). Environment, development, market, interview with Weber. *J., Natures Sciences Sociétés, 2*(3), 258265.
51. Sachs, I., (2003). Development: A key idea for the 21st century. In: Comeliau, C., (ed.), *Drafts on the Future* (Vol. 14, pp. 169–173). Contributions to the debate on alternatives, New IUED notebooks.
52. Schumacher, E. F., (1973). *Small is Beautiful: Economics as if People Mattered* (Vol. 1, pp. 5–17). London, Blond and Briggs.
53. Solow, R. M., (1992). An almost practical step toward sustainability. In: Oates, W. E., (ed.), *The RFF Reader in Environmental and Resource Management* (Vol. 1, pp. 263–272). Washington DC, Resources for the Future- 1999.
54. Stiglitz, J. E. (1970). Non-substitution theorems with durable capital goods. *The Review of Economic Studies, 37*(4), 543-553.
55. Vatin, F., (1998). *Political Economy and Natural Economy at Antoine-Augustin Cournot* (Vol. 1, pp. 1–22). Paris, PUF.
56. Vivien, F. D., (1994). *Economy and Ecology* (Vol. 1, pp. 5–27). Paris, Ed. La Découverte.
57. Vivien, F. D., (2003). *Milestones for a History of the Concept of Sustainable Development* (Vol. 31, pp. 121, 120). Mondes en Développement.
58. Vivien, F. D., (2004). *An Overview of Economic Proposals in Terms of Sustainability* (Vol. 5, No. 2, pp. 11–22). VertigO.

59. WCED, (1987). *Our Common Future: Report of the World Commission on Environment and Development* (Vol. 1, pp. 5–17). UN General Assembly, doc. A/42/427.
60. Zaccaï, E., (2002). *Sustainable Development: Dynamics and Institution of a Project* (Vol. 1, pp. 11–19). Bruxelles, P.I.E.-Peter Lang.

CHAPTER 6

Social Aspects of Sustainable Development

CONTENTS

6.1. Introduction 160

6.2. Sustainable Development (SD) Indicators 161

6.3. Social Sustainability 164

6.4. Social Aspect In Sustainable Development (SD) Of An Enterprise 166

References 171

6.1. INTRODUCTION

For development and also for environmental protection, the idea of sustainable development (SD) is contemplated as a landmark in terms of a global approach. It is termed as "development that is beneficial in meeting the requirements of the present, not jeopardizing the capability of upcoming generation to meet their requirements by themselves" (Klang et al., 2003; Vos, 2008). SD's idea involves the importance of reflecting quality that is desired, the importance of stabilizing social values and also the pace of development.

By sustainability, we mean economic growth's adjustment. Here, sustainability is referred to as the factor that is in the interest of social development and environmental protection. In this reference, to reach SD, economic sustainability is not an adequate condition but still, it is a necessary one (Roseland, 2000; Sikdar, 2003).

Everlasting economic growth that may not rely on the overuse of natural resources, or to those consequences that may raise questions on its sustainability, like the ones that are environmentally unacceptable or those consequences that may affect future generation's economic prospects, such economic growth shall be provided by SD. In terms of sustainability, Serbian economic development can be only observed on the foundation of attaining economic growth. This attained economic growth should be on the merits which shall include people, their information, education, knowledge, and the way people and institutions are connected (Callens and Tyteca, 1999; Bijl, 2011). The social aspects of sustainability totally depend on understanding people's interdependence and equality; these two are essential conditions for quality life, which is surely the utmost priority of development. For durable SD, resources, opportunities, and wealth must be divided among all citizens in a way that all citizens experience the basic security standard, the basic human rights and also the social benefits that include food, education, health, and other personality development opportunities (Kagawa, 2007; Dempsey et al., 2011). To attain SD, the essential things to take into account are (Rondinelli and Berry, 2000; Moon, 2007):

- New social values need to be developed and protected. Social capital rights shall be increased too.
- Equality for everyone in terms of security and law, regardless of gender and age, everyone shall be offered special attention.
- Safety and comprehensive health to people should be provided and promoted.

- For the preservation of identity and for the development of an individual, quality education is necessary, therefore, it should be provided.
- Solidarity among all, within a generation, between generations, solidarity with poor, with Invisibles and with marginalized groups.

At the local level and also at a national level, certain activities regarding SD were initiated by Serbia. The main challenge in many fields is local government. These fields include infrastructure, environmental protection, economic development, pervert reduction, social development, economic development, and municipal services (Lehtonen, 2004; Vavik and Keitsch, 2010).

6.2. SUSTAINABLE DEVELOPMENT (SD) INDICATORS

Indicators are qualitative guides in evaluating conditions and subsystem changes. These performance indicators are defined and developed according to the vision that is already established, policy, and also the objectives of SD (Golusin and Ivanović, 2009; Steurer and Hametner, 2013). Performances that may indicate the degree to which the established objectives are fulfilled, such degrees should be referred by these indicators. Following are the steps through which the definition and development of indicators are performed (Garcia et al., 2000; Cornelissen et al., 2001):

- The indicator's selection criteria should be defined;
- Establishing those indicators which are potentially strong;
- Valuing those indicators which are potentially strong;
- Selecting final indicators.

For allowing the supervision the of level achieved, for sustainability index and for showing the indicator's value of SD, till today a lot of models have been developed.

Following principles should be carefully considered so that the representative data can be ensured:

- Every time, the same definition should be initiated, i.e., unit measure;
- For the determination of values of indicators, same methods shall be used;

- Within the same time period, movement of indicators should be followed;
- For methods such as quality control and calculation, the same procedures should be used.

To get a better idea of the happenings and also of the things expected to happen in the upcoming time, we should focus more on indicators we have observed and their possibilities of movement trends. What will be the influence if one field among them is changed? Do they really change? Or in what orientation they move? These questions give us a better understanding of the future. Initially, a connection between groups of factors that are not connected, like the ones, environment, and economy should be discovered. Recommending the improvements and showing the present condition, through indicators projecting an upcoming period, a primary report on conditions of SD is focused. This type of report serves nothing less than a warning; it predicts the possibilities of the things that can occur, if recommended measures are not considered seriously. Indicators of regional communities (comprising local people) can be compared to the developed indicators; they also permit their comparison across the border. The indicators for SD used at the international level should be compared to the ones used at the national level. In the last 20 years, to measure the level that SD has attained, countries have started using own indicators and these countries have developed their own methodologies too. The results are being published by these countries regularly and better development is planned through National strategies. These countries include Germany, Great Britain, Belgium Sweden, and Switzerland (Azapagic and Perdan, 2000; Vera and Langlois, 2007).

For the purpose of SD, national indicators used by countries where the possibility of comparative analysis existed, lies under the minimum limit of 12 and with a max of 187 indicators (Table 6.1). For the identification of sameness between countries in regard to indicators which are common and also for issues which are of a wider perspective, specifying common indicators, 22 European countries (among 22 countries, two were situated outside Europe, i.e., Australia, and Canada) performed an analysis. It was a huge challenge to determine indexes from one country to another. It was difficult because there exist different ways of expressing indicators (Table 6.1). Social objectives and classifications in sets of indicators are identified in Table 6.3 (Victor, 1991; Mitchell et al., 1995) (Tables 6.1–6.3).

Table 6.1: Total Number of Sustainable Development's National Indicators in European Countries

Countries	Number of Indicators	Source
Germany	28	2006 Indicator Report
Greece	70	2006 Report of Sustainability Indicators
United Kingdom	147	2006 Indicator Report
Belgium	45	2008 Federal Report
France	12	National Strategy for SD
Slovenia	71	2006 Development Reports
Austria	95	The decision of the Ministerial Council

Table 6.2: Sets of Indicators Utilized Most Frequently

Number	Sustainable Development's Features	Total Sets of Indicators Found
1.	Innovation, research, and development	15
2.	Sustainable development in Global dimension	16
3.	Proper management	16
4.	Transport	16
5.	Socio-economic development	18
6.	Education	19
7.	Social issues	19
8.	Public health	19
9.	Sustainable production and consumption	20
10.	Climate changes and energy	21
11.	Natural resources management	24

Table 6.3: Social Classifications and Objectives in Social Indicator Sets

Author	Social Classification
OECD Social Indicators (OECD, 2009)	· Social "organizing dimension" Economic self-sufficiency and social cohesion · Health and equity

EU Sustainable Development Indicators (Eurostat, 2007)	· "Themes" · Social inclusion · Public health · Demography · Good governance
UN Commission for Sustainable Development (UNDESA, 2007)	· "Themes" poverty · Governance · Health education · Demographics
UN Commission for Sustainable Development (UNDESA, 2001)	· Social "themes" · Equity · Health · Education · Housing · Security (combating crime) · Population
UN Commission for Sustainable Development (UNCSD, 1996)	· Social "categories" · Combating poverty · Sustainable demographic dynamics · Human health protection · Encouraging human settlement · Encouraging public awareness, education, and training

6.3. SOCIAL SUSTAINABILITY

We can describe social sustainability as managing and setting out negative as well as positive effects of organizations, processes, systems, people's social life, and their activities. Topics like social equity and health, rights of labor, rights of human, decent, and practicing working conditions, well-being, and development of the community, cultural competence, responsibilities of product, justice, and responsibilities that are socially assigned are integrated by the concept of sustainability. There are many frameworks that are conceptually based; still, it is hard to quantify the changes. It becomes difficult to measure accuracy in changing and methodologies for evaluating it on a social basis, remains immature because of complexity and sometimes dynamic social, anthropological, and physiological factors. This might be the reason that this critically important issue gets less attention from decision-makers and from researchers as well, as compared to environmental issues and economics (Rennings and Wiggering, 1997; Callens and Tyteca, 1999).

Production chains based on biomass have a broad collection of social impacts; these impacts are on the residents, related regions, and habitats. Relying on different consideration levels of issues regarding social sustainability which include management procedures, planning, and design, the discussed impacts may induce positive as well as negative changes on the community, health, human life well-being and personal rights (Spangenberg, 2002).

Benefaction to rural development, food, and energy security, job creation, cultural as well as social acceptability, and benefaction to employment patterns are the major indicators of the performance of supply chains. Employment effect and other indicators like job creation can be observed as a socioeconomic indicator and it also serves its contribution in terms of rural development. Among all of these indicators, production chains based on biomass mostly use job creation to have an idea of social impacts. It is the most significantly computable sustainability indicator. Once the supply chains and production systems based on biomass are established, we can represent job creation by both, the quality and quantity. The quality and quantity of jobs created (it includes induced, indirect, and direct jobs) are specified by tactical decisions, operational conditions along with strategic decisions. Tactical decisions are on transportation and production level. Management, design, and planning of elements involved in a chain are different operational conditions. The location, preprocessing, storage, distribution units, the numbers and capacities, all such things related to configuration and size of the chain are included in strategic decisions. We can describe direct jobs as jobs that are created on-site or immediately, for instance, storage jobs, jobs on processing units, contractors hired, and staff hired for operations, all is direct jobs. Indirect jobs are produced as a result of increasing economic activities; these activities are increased due to the established chains. By indirect jobs, we refer to activities that take place when producers, contractors, or investors give payments to other vendors or suppliers for the goods and services they need in their business environment, for example, material suppliers, energy suppliers, accountants, and bankers. Jobs which are the result of increasing well-being and wealth are termed as induced jobs. It is the result of increasing the employment rate in the environment. By adding three different kinds of effect, one can estimate the effect on employment due to the chain units of production that is established. To assess the social impact and for systematic calculations linked with the creation of jobs, also for measuring benefits provided to employment pattern of locals, life span of supply chain units and their influence from activities like construction have to be seriously considered along with the yearly influence that operations (permanent ones) like distribution, maintenance, and production causes.

Apart from job creation, another key indicator of the production based on biomass chains is its potential to provide ease in poverty and its benefaction in the development of rural areas. Its benefaction to rural areas can be elaborated as its contribution to the well-being of the chain and environment (including residents) as well as the wealth of population. We can improve by increasing jobs opportunity which ultimately raises income per capita, fuel for agricultural and domestic use, and by supplying renewable energy (RE) resources. To compute and calculate the chain's social impact which it potentially applies on the development of the rural areas and communities living there, the generation of income and its allocation, the distribution, uses (in local people) of supply chain activities can be determined. Production chains based on biomass are usually constructed in potential areas where agricultural biomass resources can be produced. These areas are mostly rural areas. Not necessarily but these areas might be controlled by farmers maintaining their small agricultural land. Supply chain generated income, distributed rightly results in the quality life of the community, and is capable of creating equity among the community. A scheme including appealing biomass pricing for farmers, along with organized and programmed development should be done in a way that maximum advantage from such investments can be gathered (Rotheroe et al., 2003).

6.4. SOCIAL ASPECT IN SUSTAINABLE DEVELOPMENT (SD) OF AN ENTERPRISE

SD strategy has two provoking circumstances in the context of preparation. The first refers to data that is available but is outdated or collected long ago or at the republic level is rarely available. The other refers to an insufficient approach towards multidisciplinary research. This includes the problems connecting protection sectors (of the environment), education, poverty-related problems, and health. Relevant institutions working practice has shown that cross-department cooperation also lacks on all the levels, resulting in an individualistic resolution to problems that are mutually connected (Bramley and Power, 2009; Boström, 2012).

A total of 21 medium, 10 large, and 52 small-scaled organizations, altogether 83 organizations are targeted in this chapter, all from Central Serbia region where all cities have alike attributes in with regard to the living environment in present developmental conditions, social conditions, and economic development. To divide the organization on the basis of the total number of employees used, EU criteria were used which is as follow (Griessler and Littig, 2005; Dempsey et al., 2011):

- Small companies: 1–50;
- Medium companies: 51–250;
- Large companies: >250.

Mostly companies which were examined, almost 82% of them, lies under the private ownership, almost 10.8% lies under the ownership of country (or mixed) and 7.2%, lies under different sectors. Relating to different levels of SD, researchers performed research comprising a questionnaire which included 50 questions (Zhao et al., 2012; Grabara et al., 2015). The analysis was made on the literature that already existed; the empirical evaluation was performed on significant attributes. Other foreign and domestic question papers of similar interests were consulted too (Figure 6.1). For this purpose, the social conditions of all companies doing similar business activity were analyzed in the following aspects (Magis, 2010; Vallance et al., 2011):

- Per capita gross national product (C1);
- Per employee gross national product (C2);
- Mobbing (C3);
- Total occupational injuries (C4);
- Sick leave (C5);
- Training of personnel (C6);
- Qualification structure (C7);
- Women participation in the employment structure (C8);
- The age structure (C9);
- Employee satisfaction (C10).

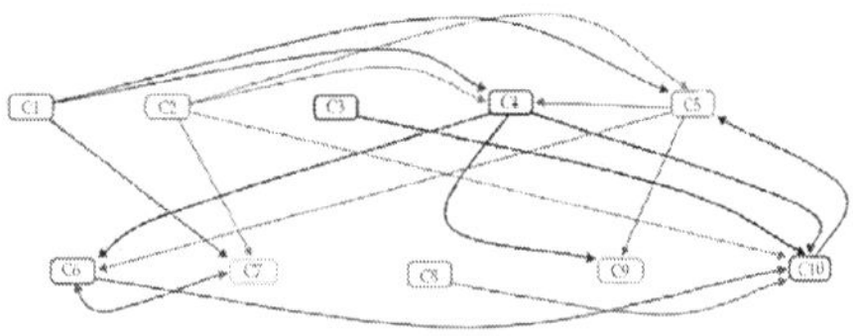

Figure 6.1: Sustainable development-social aspects model.

Source: https://www.researchgate.net/publication/321886292_SOCIAL_ASPECTS_OF_SUSTAINABLE_DEVELOPMENT_OF_ENTERPRISES.

Considering SD's social aspects in Central Serbia, compared to past times, in the recent periods of time, the number of days people pass on sick leave has decreased, though health at work and safety implementation is still ineffective. Salary cuts, job loss fear along with the economic crisis and rise in the unemployment rate are the main reasons for all this (Kernel, 2005; Rahdari et al., 2016). It is quite difficult to access the data (sick leave days). The organizations may have different methods to monitor the data, which may result in inconsistent and varying data. 3.5% of people are observed to be on leave due to sickness, illness or injury in the EU amounts whereas, in researches, it is claimed to be 6%. When we look into it, the only important thing that can be pointed out is the cost issue of these employees who are on sick leave. The cost paid to the workers who are called as replacements of originally working employees is additional to the cost that is paid to compensate the sick leaves employee. Therefore, it is an essential issue. Serious steps are being taken for the betterment of both, health, and safety issues at work. Precautionary measures are also being established so that diseases and work-related injuries can be reduced, it will help in the betterment of cooperation and social dialogues, in expanding regulatory concerns, to continue the tasks related to legislation amendment and also for inspection reinforcement in labor area (Hutchins and Sutherland, 2008).

According to the data collected in 2015, the Republic of Serbia has an average of 375 employees per 1000 residents; Central Serbia has a better average of 380 employees per 1000 residents (Sarkis et al., 2010).

With respect to the age of employees, a very small number is over the age of 65 years and similarly, very small number lies under the age of 20 years, mostly employee's fall in the age of 36 and 50 years, almost this makes 43% of the total (Figure 6.2). The structure of qualification depicts that employees having secondary level education are more in number, almost 57% whereas the unskilled and semi-qualified employees have a percentage of less than 20% (Figure 6.3).

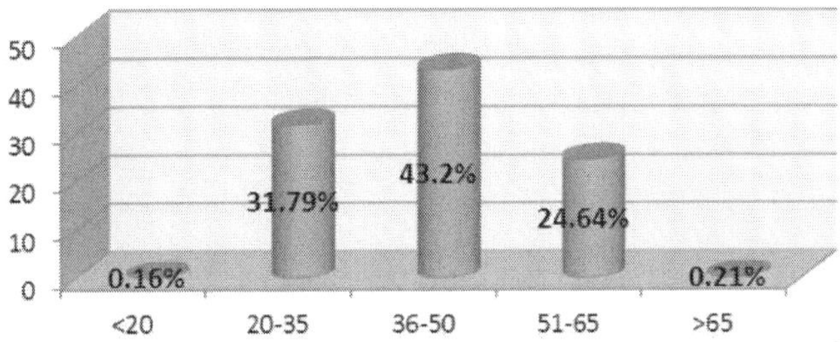

Figure 6.2: Employees age structure.

Source: https://www.researchgate.net/publication/321886292_SOCIAL_ASPECTS_OF_SUSTAINABLE_DEVELOPMENT_OF_ENTERPRISES.

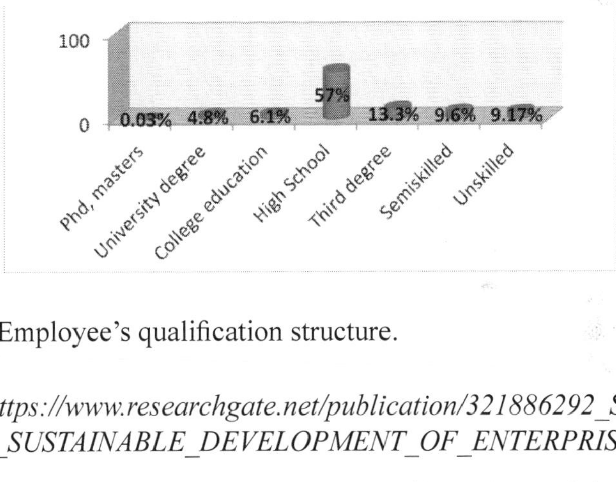

Figure 6.3: Employee's qualification structure.

Source: https://www.researchgate.net/publication/321886292_SOCIAL_ASPECTS_OF_SUSTAINABLE_DEVELOPMENT_OF_ENTERPRISES.

Throughout the last ten years, for compiling the social policies, the European efforts have boosted up for unemployed ones, specifically for women employment, people who are unemployed from a long period of time, also for employees who are old, taking special steps for minority groups and disabled ones. Legal means are encouraged when men and women are equally treated, but this will require the need for proactive intervention (Dvořáková and Zborková, 2014; Garbie, 2014). Although the law regulates men's and women's equality, the women according to mutually conditioned indicators in Serbia face extreme gender inequality issues in the labor market. These issues are comparatively of a much higher degree than compared to the neighboring countries, particularly with the European countries. There are only 22% of women, who are managerial employees whereas, in regard to the research level, the education level of men is equal to the education level of women. In addition, the number of men employees is nearly third more than the number of women employees. On a scale, with a minimum of 1 and a maximum of 10, the employee satisfaction score is almost 7 in the investigated companies, while the mobbing concerning cases are less than 0.5%, which is surely negligible (Kolk and Van Tulder, 2010; Wielgórka, 2016).

REFERENCES

1. Azapagic, A., & Perdan, S., (2000). Indicators of sustainable development for industry: A general framework. *Process Safety and Environmental Protection*, *78*(4), 243–261.
2. Bijl, R., (2011). Never waste a good crisis: Towards social sustainable development. *Social Indicators Research*, *102*(1), 157–168.
3. Boström, M., (2012). A missing pillar? Challenges in theorizing and practicing social sustainability: Introduction to the special issue. *Sustainability: Science, Practice and Policy*, *8*(1), 3–14.
4. Bramley, G., & Power, S., (2009). Urban form and social sustainability: The role of density and housing type. *Environment and Planning B: Planning and Design*, *36*(1), 30–48.
5. Callens, I., & Tyteca, D., (1999). Towards indicators of sustainable development for firms: A productive efficiency perspective. *Ecological Economics*, *28*(1), 41–53.
6. Cornelissen, A. M. G., Van Den Berg, J., Koops, W. J., Grossman, M., & Udo, H. M., (2001). Assessment of the contribution of sustainability indicators to sustainable development: A novel approach using fuzzy set theory. *Agriculture, Ecosystems and Environment*, *86*(2), 173–185.
7. Dempsey, N., Bramley, G., Power, S., & Brown, C., (2011). The social dimension of sustainable development: Defining urban social sustainability. *Sustainable Development*, *19*(5), 289–300.
8. Dvořáková, L., & Zborková, J., (2014). Integration of sustainable development at enterprise level. *Procedia Engineering*, *69*, 686–695.
9. Garbie, I. H., (2014). An analytical technique to model and assess sustainable development index in manufacturing enterprises. *International Journal of Production Research*, *52*(16), 4876–4915.
10. Garcia, S. M., Staples, D. J., & Chesson, J., (2000). The FAO guidelines for the development and use of indicators for sustainable development of marine capture fisheries and an Australian example of their application. *Ocean and Coastal Management*, *43*(7), 537–556.
11. Golusin, M., & Ivanović, O. M., (2009). Definition, characteristics, and state of the indicators of sustainable development in countries of Southeastern Europe. *Agriculture, Ecosystems and Environment*,

130(1/2), 67–74.

12. Grabara, J., Bajdor, P., & Mihaescu, L., (2015). Steps of sustainable development implementation into enterprise activities. *Management of Sustainable Development*, *7*(1), 45–49.
13. Griessler, E., & Littig, B., (2005). Social sustainability: A catchword between political pragmatism and social theory. *International Journal for Sustainable Development*, *8*(1/2), 65–79.
14. Hutchins, M. J., & Sutherland, J. W., (2008). An exploration of measures of social sustainability and their application to supply chain decisions. *Journal of Cleaner Production*, *16*(15), 1688–1698.
15. Kagawa, F., (2007). Dissonance in students' perceptions of sustainable development and sustainability: Implications for curriculum change. *International Journal of Sustainability in Higher Education*, *8*(3), 317–338.
16. Kernel, P., (2005). Creating and implementing a model for sustainable development in tourism enterprises. *Journal of Cleaner Production*, *13*(2), 151–164.
17. Klang, A., Vikman, P. Å., & Brattebø, H., (2003). Sustainable management of demolition waste: An integrated model for the evaluation of environmental, economic and social aspects. *Resources, Conservation and Recycling*, *38*(4), 317–334.
18. Kolk, A., & Van Tulder, R., (2010). International business, corporate social responsibility, and sustainable development. *International Business Review*, *19*(2), 119–125.
19. Lehtonen, M., (2004). The environmental-social interface of sustainable development: Capabilities, social capital, institutions. *Ecological Economics*, *49*(2), 199–214.
20. Magis, K., (2010). Community resilience: An indicator of social sustainability. *Society and Natural Resources*, *23*(5), 401–416.
21. Mitchell, G., May, A., & McDonald, A., (1995). PICABUE: A methodological framework for the development of indicators of sustainable development. *The International Journal of Sustainable Development and World Ecology*, *2*(2), 104–123.
22. Moon, J., (2007). The contribution of corporate social responsibility to sustainable development. *Sustainable Development*, *15*(5), 296–306.
23. Murphy, K., (2012). The social pillar of sustainable development: A literature review and framework for policy analysis. *Sustainability:*

Science, Practice and Policy, *8*(1), 15–29.

24. Rahdari, A., Sepasi, S., & Moradi, M., (2016). Achieving sustainability through Schumpeterian social entrepreneurship: The role of social enterprises. *Journal of Cleaner Production*, *137*, 347–360.
25. Rennings, K., & Wiggering, H., (1997). Steps towards indicators of sustainable development: Linking economic and ecological concepts. *Ecological Economics*, *20*(1), 25–36.
26. Rondinelli, D. A., & Berry, M. A., (2000). Environmental citizenship in multinational corporations: Social responsibility and sustainable development. *European Management Journal*, *18*(1), 70–84.
27. Roseland, M., (2000). Sustainable community development: Integrating environmental, economic, and social objectives. *Progress in Planning*, *54*(2), 73–132.
28. Rotheroe, N., Keenlyside, M., & Coates, L., (2003). Local agenda 21, articulating the meaning of sustainable development at the level of the individual enterprise. *Journal of Cleaner Production*, *11*(5), 537–548.
29. Sarkis, J., Helms, M. M., & Hervani, A. A., (2010). Reverse logistics and social sustainability. *Corporate Social Responsibility and Environmental Management*, *17*(6), 337–354.
30. Sikdar, S. K., (2003). Sustainable development and sustainability metrics. *AIChE Journal*, *49*(8), 1928–1932.
31. Spangenberg, J. H., (2002). Environmental space and the prism of sustainability: Frameworks for indicators measuring sustainable development. *Ecological Indicators*, *2*(3), 295–309.
32. Steurer, R., & Hametner, M., (2013). Objectives and indicators in sustainable development strategies: Similarities and variances across Europe. *Sustainable Development*, *21*(4), 224–241.
33. Vallance, S., Perkins, H. C., & Dixon, J. E., (2011). What is social sustainability? A clarification of concepts. *Geoforum*, *42*(3), 342–348.
34. Vavik, T., & Keitsch, M. M., (2010). Exploring relationships between universal design and social sustainable development: Some methodological aspects to the debate on the sciences of sustainability. *Sustainable Development*, *18*(5), 295–305.
35. Vera, I., & Langlois, L., (2007). Energy indicators for sustainable development. *Energy*, *32*(6), 875–882.
36. Victor, P. A., (1991). Indicators of sustainable development: Some lessons from capital theory. *Ecological Economics*, *4*(3), 191–213.

37. Vos, R., (2008). Defining sustainability: A conceptual orientation. *Journal of Chemical Technology and Biotechnology, 8*, pp 334–339.
38. Wielgórka, D., (2016). Environmental management in the aspect of sustainable development in micro-, small-, and medium-sized enterprises. *Desalination and Water Treatment, 57*(3), 982–992.
39. Zhao, Z. Y., Zhao, X. J., Davidson, K., & Zuo, J., (2012). A corporate social responsibility indicator system for construction enterprises. *Journal of Cleaner Production, 29*, 277–289.

CHAPTER 7

Future Prospects of Sustainable Development

CONTENTS

7.1. Introduction 176

7.2. Human Well-Being And Competencies 177

7.3. Sustainable And Economies 178

7.4. Food Systems And The Patterns Of Nutrition 179

7.5. Energy De-Carbonization With Worldwide Access 180

7.6. Urban And Peri-Urban Development 181

7.7. Global Environmental Commons 182

7.8. Science For Ecological Development 183

7.9. Not Incremental Alteration And Transformation 183

References 185

7.1. INTRODUCTION

The latest Report of Global Sustainable Development (SD) recognizes s6 entry points that deliver the most potential for accomplishing the anticipated transformations at the necessary scale and the speed. In doing so the report considers the earnestness, the forward-looking potentials about a increasing global population looking for higher levels of well-being and normative considerations, like leaving behind no one, these aren't entry points into specific or even groups of Goals, but instead into the fundamental systems (Haggag and Ayad, 2002; Ehnert and Harry, 2012). Simultaneously, not attending to inter-linkages that are inherent to the entry points and cut across them-for instance, through concentrating on specific targets and goals-would risk progress across several elements of 2030 Agenda. The nominated entry points are:

- Human well-being and capabilities;
- Food systems and nutrition patterns;
- Sustainable and just economies;
- Energy de-carbonization with universal access;
- Global environmental commons;
- Urban and pre-urban development.

The Report also recognizes four levers, which can be deployed coherently through every entry point in order to bring about the required transformations:

- Governance;
- Individual and collective action;
- Economy and finance; and
- Science and technology.

The levers are associated with means of execution described in Goal 17, but are different, in that they assist the multiple, corresponding roles that specific entities and actors play in bringing the change. Every lever can back individually to the systemic change; though, the chapter discusses that it is through their context-dependent amalgamations that it will be feasible to bring the transformations essential for corresponding across the dimensions of ecological development and accomplishing the 2030 Agenda (Adamantiades and Kessides, 2009; Whitehead, 2012).

Decision-makers must act based on understanding and current knowledge of the related human-social-environmental systems at any level.

The knowledge also necessities to be extensively accessible to all actors and countries, encouraging innovative partnerships and coalitions for success (Williams et al., 2011; Samsudin et al., 2016).

Moreover, new technological and scientific research, along with the adaptation of prevailing technologies and knowledge to particular regional and local contexts, are required to further streamline struggles, maximize synergies amongst the Goals, and pre-emptively assist evolving tests beyond the 2030 horizon. The chapter constitutes a novelty in a way the scientific expertise is organized by the system of the United Nations as a whole. It proposes new methods of strengthening the impact of technology and science to the Agenda of 2030, helping enhance the science-policy interface (Martínez-Alier et al., 2010; Larrea, 2013).

The approaches and call in order to action anticipated in the Report of United Nations for each of 6 entry points for alterations, and for enhancing the part of science in executing the Goals, are given below.

7.2. HUMAN WELL-BEING AND COMPETENCIES

Progressing human well-being-comprising material well-being, education, voice, health, access to a safe and clean environment and flexibility-is at the center of changes towards SD. Not only is the well-being of humans inherently significantly, but the capabilities of people, in turn, drive global economic, environmental, and social change according to the sets of skills, knowledge, competencies, and physical and psychological abilities. Health and education aren't just development outcomes. They are the means of accomplishing the key features of the worldwide development agenda (McGillivray and Clarke, 2006; Landorf et al., 2008).

The world has made considerable progress in the well-being of humans in recent decades, but utmost deprivations remain, and advancement remains uneven. Utmost poverty-described as living below the fiscal threshold of 1.90 dollars per person/day-was at around 8.6% of the global population in the year 2018, and is focused-with around more than half of the extremely poor of the world living in 5 countries in South Asia and sub-Saharan Africa. In the year 2030, delicate States affected by conflict and crisis will be home to around 85% of those lingering in utmost poverty (Dodds, 1997; Lessmann and Rauschmayer, 2013).

Current estimates specify that the world isn't on track, without extra struggle, to eliminate extreme poverty by the year 2030. Extreme poverty is concentrated among the marginalized groups-women, ethnic minorities, indigenous peoples, persons with disabilities and the others. Gender inequality, which restricts the capabilities and opportunities of half of the population of the world, further aggravates the women's condition in poverty. In several places, there exist socioeconomic gaps amongst persons with and the persons without disabilities, as the persons with disabilities frequently suffer lower levels of education, absence of social protection in contrast with their peers, and higher rates of economic inactivity and unemployment (Breslow et al., 2016).

Effective action in any area needs recognizing and addressing the associates among them-the close links between human health and climate change, for example, or the methods in which the biodiversity loss and decline of ecosystem services worsen inequalities. Paths to advance the well-being of humans ultimately need cooperation, association, and dialogue between multiple actors, and engaging several levers of change. There isn't any single path, and different amalgamations of efforts are needed across regions and for the countries in special circumstances (Rauschmayer et al., 2012).

7.3. SUSTAINABLE AND ECONOMIES

Economic growth has augmented national incomes considerably, albeit unevenly, across the countries. Whereas that has assisted in progress in human, economic, and social well-being, the impacts on the environment and the human societies are currently unsustainable. Economic activity must not be observed as an end in itself, but instead as a mean for sustainably progressing human capabilities. Decoupling the advantages of economic activity from the expenses at all of the levels is crucial in itself and can assist the systemic transformations envisioned through the other 5 entry points encouraged in this chapter. Such a result would greatly speed up the essential reconfiguration and assistance to put people, nature, and societies on the pathway to SD (Chouinard et al., 2011; Cock, 2014).

Currently, there are many reasons why that aren't happening. One often-cited reason is the utilization of the GDP (gross domestic product)-the market value of services and goods yielded over a year-as the only or major metric for managing economic policy for the human development. Whereas reforming the making of policy at this level is mandatory, it might not happen quickly enough across the globe to guarantee efficient paths towards SD.

Conversely, numerous other substantial impairments could be addressed. Production valuations don't account for all expenses or the value-added, as the prices charged for services and goods don't reflect the full expenses of negative externalities, like waste generated and discharged into the environment. Continually increasing the ingesting of waste-generating services and goods globally is unsustainable. On current trends, yearly global resource utilization is anticipated to reach around 18 tons per capita by the year 2060, with unsustainable influences from increases in the greenhouse gas emissions, agricultural land area, and industrial water withdrawals. Inspecting the life cycles of particular items, like electronics and plastics, leads to similar conclusions. Indeed, economic, and social deprivations in several parts of the globe can be addressed through increasing usage, but that must be balanced by shifting usage worldwide towards services and goods produced with quite lower environmental impact (Jacobs and Stott, 1992).

Investment in sustainable development goals (SDGs) from the sources is considerably short of what is required. Production across national commands also triggers its own challenges (O'Riordan and Voisey, 1997; Essen, 2010). Whereas globalization has backed to generating jobs, reducing poverty, allowing greater access to a broader range of sparking innovation and products, the scattering of production across diverse national controls can also outcome in the race to bottom in terms of labor and environmental standards. Nationally concluded instruments, like taxes or regulations, might not be sufficient to manage the effects (Bidegain Ponte and Enríquez, 2016).

7.4. FOOD SYSTEMS AND THE PATTERNS OF NUTRITION

Food is mandatory for survival of humans, and its providing employs over 1 billion people. The worldwide food system comprises several regional and local food systems. It comprises not only the production of food but also all of the food-associated activities and how the activities intermingle with the natural resources and processes of Earth. Due to its climate and environmental shortcomings and impacts in safe, healthy nutrition for all, the global food system of today is unsustainable. Moreover, it doesn't confirm healthy food patterns for the population of world. It is approximated that around 820 million people in the world are still hungry. Simultaneously, overweight, and rising obesity can be observed in every region around the globe. Globally, 2 billion adults are generally overweight, as are 40 million children (Smith, 2013; Gillespie and van den Bold, 2017).

Billions of land hectares have been degraded already, and a surplus 12 million hectares of land is expected to become useless for the production of food every year. Moreover, agricultural practices can trigger the eutrophication of aquatic environment, soil acidification, atmospheric pollution, and groundwater contamination. Those practices were responsible for 60% of the worldwide emissions of greenhouse gas nitrous oxide in 2011. However, the share of nitrous oxide from agriculture appears to be decreasing. When all emissions linked with the worldwide food system are well-thought-out, they account for around 19–29% of the total greenhouse gas emissions. Without any technological developments or some other forms of mitigation, particularly the repair of soil health to increase its content of carbon, greenhouse gas emissions from worldwide agriculture could increase by 87% if production is increased in order to meet the demands of the global population in the year 2050. That scenario is mismatched with the Paris Agreement and SD Agenda (Pingali, 2007; Lv and Brown, 2010).

Scaling up the system of food as it prevails today in order to feed the rising world's population through the year 2050 and beyond, whereas accepting non-food agricultural supplies is an overarching apprehension. However, under the business-as-usual circumstances, projected 637 million people of the world will be malnourished and the environmental influences of increased production would eradicate any chance of accomplishing the Goals of 2030 Agenda. Furthermore, crop diseases and pests put worldwide food supplies at danger; but managing the supplies with increased utilization of the chemical inputs could endanger numerous environment-associated Goals (Kuhnlein and Receveur, 1996; Kennedy et al., 2004).

7.5. ENERGY DE-CARBONIZATION WITH WORLDWIDE ACCESS

Access to energy is recognized universally as a key to the economic development and realization of social and human well-being. Energy poverty remains widespread, with close to 1 billion people without any access to electricity-mainly in sub-Saharan Africa—and around 3 billion people depending on contaminating solid fuels for the purpose of cooking, which triggers a projected 3.8 million premature demises every year, according to WHO (World Health Organization). In several regions, the current utilization of biomass fuels needs children and women to spend several hours per week gathering and carrying customary biomass that is usually burned in highly polluting and inefficient stoves. Yet, generation of electricity, production of heat and transport depend mainly on the fossil fuels and together make up roughly 70% of the global greenhouse gas emissions, comprising 40% from electricity (Steinberg, 1999; Di Silvestre et al., 2018). The fastest advancement in renewables lingers to be in the generation of electricity, were close to 25% came from renewables in the year 2016, thanks to the quick expansion of the solar photovoltaic and wind. The utilization of modern renewables for transport and heat remains restricted, with shares of 9% and 3.3%, respectively. Considering that transport and heat represent 80% of the total final energy utilization, particular efforts are required in the areas to speed up the uptake of renewables. With renewable energy (RE) gradually governing power production, upgrading of transport and distribution of electricity, comprising options like hydrogen and the storage technologies and electrification of the energy end usages can become the drivers of de-carbonization in the sector of energy (Goldthau, 2011; Alstone et al., 2015).

Technologies already prevail for moving to de-carbonized paths. In the year 2016, around one-fourth of a generation of electricity came from the renewables, comprising wind and solar PV. However, advancement has been hindered by slow advancement in smart-grid management and the long-term storage of electricity. The amount of contemporary RE in total worldwide energy supply upsurged by an average of 5.4% yearly over the past era and for 5 years in a row (2014–2018), worldwide investments in the clean energy surpassed 300 dollars billion annually. That was assisted by the point that, since the year 2009, the cost of renewable electricity has fallen by 77% for the solar PV and by 38% for the onshore wind, whereas the price of electricity from the conventional sources has suffered only modest reductions (Åhman et al., 2017; Deichmann et al., 2011).

7.6. URBAN AND PERI-URBAN DEVELOPMENT

If present trends carry on, cities will comprise around 70% of the population of the world and yield 85% of worldwide economic output by the year 2050. The environmental and human influence of cities is amazingly high, and enforces a high price on the surrounding rural areas. 90% of people living in the cities breathe air that generally fails to meet the WHO standards of the quality of air; no metropolitan city in Asia or sub-Saharan Africa meets that standard. The water footmark of cities-the water source area-comprises

41% of the surface of Earth, whereas the physical footprint-the land area-comprises only 2%; the land engaged by the cities in developing world will almost triple by the year 2050. Cities are accountable for 70% of the worldwide greenhouse gas emissions from the burning of fossil fuels and will normally need to become the neutral carbon if the globe is to accomplish the targets contained in Paris Agreement. If the development endures in a business-as-usual model, the cities of the world will consume around 90 billion tons each year of the raw materials, like sand, gravel, coal, iron ore, and wood, by the year 2050. That will have irreparable outcomes on the reduction of the finite resources and will mean the devastation of green space and natural habitats, and subsequent loss of biodiversity. In several cases, urbanization is continuing organically, without any planning, and as the urban centers concentrate in the coastal areas, urban inhabitants live with a risk of mudslides, flooding, and other disasters (Briggs and Mwamfupe, 2000; Yang et al., 2010).

7.7. GLOBAL ENVIRONMENTAL COMMONS

The worldwide environmental commons include the atmosphere, the global ocean, the hydrosphere, the cryosphere, large-scale biomes, Polar Regions, and the natural resources systems like forests, land, biodiversity, and water, which make up the shared resources of Earth. The commons assist to the working of biosphere-the worldwide ecological system—and are important for survival and well-being of humans. Conditions on the Earth are shaped by interaction among all of the living organisms and the climate system. Consequently, variations in the functioning of the biosphere caused by the activities of humans are ultimately reflected in the global environmental conditions on the Earth (Anderson, 1989; Pinto and De Oliveira, 2008).

Confirming the long-term health of worldwide environmental commons is thus essential. The current action of humans is quickly depleting and humiliating the commons. There is a crucial need to manage that how the resources are extracted from commons, how proficiently the resources are utilized, how they are dispersed, and the how waste is disposed of. Since the worldwide environmental commons are fundamentally linked to each other, accomplishing sustainability of the systems of Earth needs expecting feedback impacts among the commons to maximize the co-benefits and lessen trade-offs, both locally and globally (Boda, 2003).

Breaking the limits of these systems presents dangers that experience severe economic, political, and social consequences. In Summary for the policymakers of the worldwide assessment report on ecosystem and biodiversity services, the IPBES (Intergovernmental Science-Policy Platform on Biodiversity and Ecosystem Services) described that nature across the world has been considerably transformed by several human drivers, with the maximum of indicators of biodiversity and ecosystems showing a quick decline. 75% of the land surface of the Earth has been considerably altered, 66% of the area of the ocean is suffering increasing cumulative influences, and over 85% of the wetlands have been lost (O'Connor, 2008).

7.8. SCIENCE FOR ECOLOGICAL DEVELOPMENT

For worse or for better, science, and technology are very powerful agents of transformation, depending on how they both are steered. Guided by 2030 Agenda, upsurged science-policy-society collaboration can attach breakthroughs in the understanding of joined human-environment systems and shaping of the innovative paths towards accomplishing the SDGs. The point that many countries are now including science, innovation, and technology in their national growth agenda is an encouraging sign (Funtowicz et al., 1998; Verburg et al., 2015).

Quick technological progress in computer sciences, biotechnologies, and artificial intelligence hold the promise of offering solutions to several of the problems facing the SDGs, comprising those that include difficult trade-offs.

Simultaneously, technological novelties risk further entrenching prevailing inequalities, presenting new ones and through unintended concerns, setting back advancement towards the 2030 Agenda. For instance, without any access to the digital infrastructure and available information and the communication technology, individuals with disabilities are at augmented danger of being excepted from surveys and statistics utilized to develop future policies and programs (Kläy et al., 2015).

7.9. NOT INCREMENTAL ALTERATION AND TRANSFORMATION

The 2030 Agenda is something more than a sum of the measurable Goals, indicators, and targets. It is a normative direction and the escort for action for recognizing and following SD urgencies and creating consistency between sectors and policies, in all of the contexts-local, national, regional, transnational, and worldwide (Roggema et al., 2012; Borel-Saladin and Turok, 2013). Whereas the six entry points and four levers anticipated in the report of Global SD specify a plan of action, they don't offer an exhaustive coverage of problems to accomplishing the 2030 Agenda. The six entry points and four levers should instead be utilized as references in order to guide the countries and all of the actors in their own context-particular execution strategies for accomplishing the SDGs (Park et al., 2012; Norris and Reddick, 2013).

REFERENCES

1. Adamantiades, A., & Kessides, I., (2009). Nuclear power for sustainable development: Current status and future prospects. *Energy Policy*, *37*(12), 5149–5166.
2. Åhman, M., Nilsson, L. J., & Johansson, B., (2017). Global climate policy and deep decarbonization of energy-intensive industries. *Climate Policy*, *17*(5), 634–649.
3. Alstone, P., Gershenson, D., & Kammen, D. M., (2015). Decentralized energy systems for clean electricity access. *Nature Climate Change*, *5*(4), 305–314.
4. Anderson, F. R., (1989). Of Herdsmen and Nation States: The global environmental commons. *Am. UJ Int'l L., & Pol'y*, *5*, 217.
5. Banzhaf, E., Grescho, V., & Kindler, A., (2009). Monitoring urban to peri-urban development with integrated remote sensing and GIS information: A Leipzig, Germany case study. *International Journal of Remote Sensing*, *30*(7), 1675–1696.
6. Bidegain Ponte, N., & Enríquez, C. R., (2016). Agenda 2030: A bold enough framework towards sustainable, gender-just development? *Gender and Development*, *24*(1), 83–98.
7. Boda, Z., (2003). Global environmental commons and the need for ethics. *Society and Economy, in Central and Eastern Europe | Journal of the Corvinus University of Budapest*, *25*(2), 213–224.
8. Borel-Saladin, J. M., & Turok, I. N., (2013). The green economy: Incremental change or transformation? *Environmental Policy and Governance*, *23*(4), 209–220.
9. Breslow, S. J., Sojka, B., Barnea, R., Basurto, X., Carothers, C., Charnley, S., & Hicks, C. C., (2016). Conceptualizing and operationalizing human wellbeing for ecosystem assessment and management. *Environmental Science and Policy*, *66*, 250–259.
10. Briggs, J., & Mwamfupe, D., (2000). Peri-urban development in an era of structural adjustment in Africa: the city of Dares Salaam, Tanzania. *Urban Studies*, *37*(4), 797–809.
11. Chouinard, Y., Ellison, J., & Ridgeway, R., (2011). The sustainable economy. *Harvard Business Review*, *89*(10), 52–62.
12. Cock, J., (2014). The 'green economy': A just and sustainable development path or a 'wolf in sheep's clothing'? *Global Labor Journal*, *5*(1), 1–20.

13. Deichmann, U., Meisner, C., Murray, S., & Wheeler, D., (2011). The economics of renewable energy expansion in rural Sub-Saharan Africa. *Energy Policy*, *39*(1), 215–227.
14. Di Fabio, A., & Rosen, M. A., (2018). Opening the black box of psychological processes in the science of sustainable development: A new frontier. *European Journal of Sustainable Development Research*, *2*(4), 47.
15. Di Silvestre, M. L., Favuzza, S., Sanseverino, E. R., & Zizzo, G., (2018). How decarbonization, digitalization and decentralization are changing key power infrastructures. *Renewable and Sustainable Energy Reviews*, *93*, 483–498.
16. Dodds, S., (1997). Towards a 'science of sustainability': Improving the way ecological economics understands human well-being. *Ecological Economics*, *23*(2), 95–111.
17. Ehnert, I., & Harry, W., (2012). Recent developments and future prospects on sustainable human resource management: Introduction to the special issue. *Management Revue*, *1*, 221–238.
18. Essen, J. M., (2010). Sufficiency economy and Santi Asoke: Buddhist economic ethics for a just and sustainable world. *Journal of Buddhist Ethics*, *17*, 69–99.
19. Funtowicz, S., Ravetz, J., & O'Connor, M., (1998). Challenges in the use of science for sustainable development. *International Journal of Sustainable Development*, *1*(1), 99–107.
20. Gillespie, S., & van den Bold, M., (2017). Agriculture, food systems, and nutrition: Meeting the challenge. *Global Challenges*, *1*(3), 1600002.
21. Goldthau, A., (2011). Governing global energy: Existing approaches and discourses. *Current Opinion in Environmental Sustainability*, *3*(4), 213–217.
22. Haggag, M. A., & Ayad, H. M., (2002). The urban structural units method: A basis for evaluating environmental prospects for sustainable development. *Urban Design International*, *7*(2), 97–108.
23. Jacobs, M., & Stott, M., (1992). Sustainable development and the local economy. *Local Economy*, *7*(3), 261–272.
24. Kennedy, G., Nantel, G., & Shetty, P., (2004). Globalization of food systems in developing countries: A synthesis of country case studies. *Globalization of Food Systems in Developing Countries: Impact on*

Food Security and Nutrition, *83*(1), 11–30.

25. Kläy, A., Zimmermann, A. B., & Schneider, F., (2015). Rethinking science for sustainable development: Reflexive interaction for a paradigm transformation. *Futures*, *65*, 72–85.
26. Kuhnlein, H. V., & Receveur, O., (1996). Dietary change and traditional food systems of indigenous peoples. *Annual Review of Nutrition*, *16*(1), 417–442.
27. Landorf, H., Doscher, S., & Rocco, T., (2008). Education for sustainable human development: Towards a definition. *Theory and Research in Education*, *6*(2), 221–236.
28. Larrea, C., (2013). Extractivism, economic diversification, and prospects for sustainable development in Ecuador. In: *Latin America and Shifting Sands of Global Power Conference, Australia National University, Canberra* (Vol. 1, pp. 11–12).
29. Lessmann, O., & Rauschmayer, F., (2013). Re-conceptualizing sustainable development on the basis of the capability approach: A model and its difficulties. *Journal of Human Development and Capabilities*, *14*(1), 95–114.
30. Lv, N., & Brown, J. L., (2010). Chinese American family food systems: Impact of Western influences. *Journal of Nutrition Education and Behavior*, *42*(2), 106–114.
31. Mandere, M. N., Ness, B., & Anderberg, S., (2010). Peri-urban development, livelihood change, and household income: A case study of peri-urban Nyahururu, Kenya. *Journal of Agricultural Extension and Rural Development*, *2*(5), 73–83.
32. Martínez-Alier, J., Pascual, U., Vivien, F. D., & Zaccai, E., (2010). Sustainable de-growth: Mapping the context, criticisms and future prospects of an emergent paradigm. *Ecological Economics*, *69*(9), 1741–1747.
33. McGillivray, M., & Clarke, M., (2006). Human well-being: Concepts and measures. *Understanding Human Well-Being*, *1*, 3–16.
34. Norris, D. F., & Reddick, C. G., (2013). Local e-government in the United States: Transformation or incremental change? *Public Administration Review*, *73*(1), 165–175.
35. O'Connor, D., (2008). Governing the global commons: Linking carbon sequestration and biodiversity conservation in tropical forests. *Global Environmental Change*, *18*(3), 368–374.

36. O'Riordan, T., & Voisey, H., (1997). The political economy of sustainable development. *Environmental Politics*, *6*(1), 1–23.
37. Park, S. E., Marshall, N. A., Jakku, E., Dowd, A. M., Howden, S. M., Mendham, E., & Fleming, A., (2012). Informing adaptation responses to climate change through theories of transformation. *Global Environmental Change*, *22*(1), 115–126.
38. Pingali, P., (2007). Westernization of Asian diets and the transformation of food systems: Implications for research and policy. *Food Policy*, *32*(3), 281–298.
39. Pinto, R. F., & De Oliveira, J. A. P., (2008). Implementation challenges in protecting the global environmental commons: The case of climate change policies in Brazil. *Public Administration and Development: The International Journal of Management Research and Practice*, *28*(5), 340–350.
40. Rauschmayer, F., Omann, I., & Frühmann, J., (2012). Needs, capabilities, and quality of life: refocusing sustainable development: Felix Rauschmayer, Ines Omann and Johannes Frühmann. In: *Sustainable Development* (Vol. 1, pp. 23–46). Routledge.
41. Roggema, R., Vermeend, T., & Dobbelsteen, A. V. D., (2012). Incremental change, transition, or transformation? Optimizing change pathways for climate adaptation in spatial planning. *Sustainability*, *4*(10), 2525–2549.
42. Samsudin, M. S. N., Rahman, M. M., & Wahid, M. A., (2016). Power Generation Sources in Malaysia: Status and prospects for sustainable development. *Journal of Advanced Review on Scientific Research*, *25*(1), 11–28.
43. Smith, I. F., (2013). Sustained and integrated promotion of local, traditional food systems for nutrition security. In: *Diversifying Food and Diets* (Vol. 1, pp. 154–171). Routledge.
44. Steinberg, M., (1999). Fossil fuel decarbonization technology for mitigating global warming. *International Journal of Hydrogen Energy*, *24*(8), 771–777.
45. Verburg, P. H., Crossman, N., Ellis, E. C., Heinimann, A., Hostert, P., Mertz, O., & Grau, R., (2015). Land system science and sustainable development of the earth system: A global land project perspective. *Anthropocene*, *12*, 29–41.

46. Whitehead, M., (2012). The sustainable city: An obituary? On the future form and prospects of sustainable urbanism. *The Future of Sustainable Cities: Radical Reflections*, *1*, 29–46.
47. Williams, B., Wilmshurst, T., & Clift, R., (2011). Sustainability reporting by local government in Australia: Current and future prospects. In: *Accounting Forum* (Vol. 35, No. 3, pp. 176–186). Taylor & Francis.
48. Yang, Z., Cai, J., & Sliuzas, R., (2010). Agro-tourism enterprises as a form of multi-functional urban agriculture for peri-urban development in China. *Habitat International*, *34*(4), 374–385.

CHAPTER 8

Issues in Sustainable Development and Their Potential Solutions

CONTENTS

8.1. Introduction 192

8.2. The Necessity For Sustainable Development (SD) 193

8.3. The Origin of Sustainable Development (SD) 195

8.4. Hurdles In The Path of Sustainable Development (SD) 197

8.5. Potential Solutions 202

8.6. Creative Solutions For Sustainability 204

8.7. Areas of Innovative Solutions Requiring Focus 206

8.8. Role of The United Nations 209

References 214

8.1. INTRODUCTION

Sustainable development (SD) has surfaced as a possible solution to the increasing spectrum of environmental and social challenges brought about by the uninterrupted proliferation of unsustainable patterns of production and consumption linked with the expected level of the population as a failure of developmental planning. Progress is poor, despite increasing attempts to marry the environmental and social challenges with economic growth. Hence, this chapter focuses on exploring the basis of the present unsatisfactory progress concerning the practical application of the concept. This chapter emphasizes the deficiencies of the modern societal conflict resolution mechanism: political and market arenas. A precautionary principle, which guarantees a superior decision-making process, is a possible solution that suggests switching to the third arena which promotes a wide-ranging use of innovative tools, integration of public debates, and scientific evidence and theory (McRae et al., 1989; Daniels, 1992).

SD, which is a characteristic example of the development of a new world order, introduces ways required to fulfill the demands of the current generation without sabotaging the opportunities and possibilities of future generations by attaining environmental, social, and economic unification. Having gone through all the anticipated stages, i.e., from an intriguing movement met with the same degree of reluctance, doubt, and interest to being a philosophical side-show, a plan with people having conflicting point of views while being unquestionable at the same time, this approach has been adopted by organizations, individuals, and companies as their own (Uddin et al., 2010; Hwang and Tan, 2012). Generally, it started as an unnoticeable trend, eventually taking shape to emerge as a fundamental global condition (Nordic, 2002). The significant rise of the interest in sustainability by the global media is depicted in Figure 8.1. Whereas the efforts that were being made both worldwide and in the UK to progress towards sustainability can be seen in Figure 8.2.

Well-publicized ceremonies and gatherings intended to facilitate SD are being held increasingly, however, their practical implementation has been seen to be strikingly low (NFSD, 2002; Dyllick and Hockerts, 2002). This chapter investigates the reasons behind this and potentially proposes a way forward by analyzing a large body of knowledge and inspecting the different attempts made to achieve sustainability. Moreover, in this chapter, we will also investigate an array of issues and barriers associated with sustainability (Kühtz, 2007; Urbaniec, 2015).

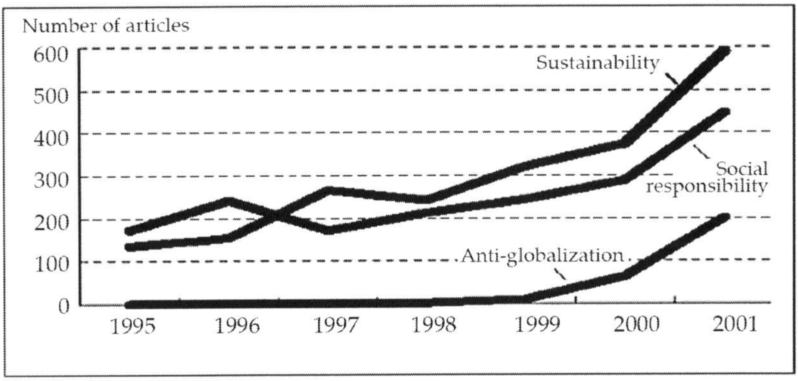

Figure 8.1: Keywords in twelve leading international media 1995–2001 (Nordic, 2002).

Source: https://www.irbnet.de/daten/iconda/CIB10669.pdf.

8.2. THE NECESSITY FOR SUSTAINABLE DEVELOPMENT (SD)

The growth of SD has been driven by an interest in ecology and population-related statistics, boosting industrial activity and value creation. An ideal society can be illustrated through sustainability which promotes high standards of living for the current and future generations (Brundtland, 1987; DETR, 1999). The relevant literature encompasses a wide range of sustainability challenges (IPCC, 1996; UNDP, 1999, 2000).

8.2.1. Demography, Economic Interest and, Poverty

Due to the rise in population by 140% during the past 50 years, the Earth is presently inhabited by 6 billion people and is expected to house 9 billion by 2025 (UN, 2002). However, one fifth out of those who reside here are deprived of medical facilities, more than 40% live in the absence of the latest sanitary and energy resources and one-sixth suffer from the unavailability of safe drinking water (DTI, 2004). While 50% continue to live on an income, lower than 2 dollars a day (Leisinger and Schmitt, 2002).

8.2.2. Standard of Living, Population Analysis, Poverty, and Urbanization

Several countries in the present statistically consist of younger people, while the planet is naturally aging (Fussler and James, 1996). The number of live births per thousand of population per year is diminishing in developed counties as the number of years a person is expected to live increases. This means that: advantages such as public health facilities and pension schemes will be unfeasible; there will be delayed economic progress. Regional inequality is one of the EU's greatest worries as 1 out of 6 Europeans live a life of insufficiency.

8.2.3. Environmental Deterioration, Value Creation, and Industrial Interest

Production of solid waste, deforestation, high levels of CO_2 discharge and decline in the quality of water and air are some of the most crucial environmental problems caused by the activities in the industrial sector today. The rate of environmental deterioration along with the exhaustion of nature's recourses calls for immediate action to dissociate economic development from environmental deterioration. To attain "Factor 10" improvement, a few have urged to withdraw from 90% of economic proceedings (Schmidt-Bleek, 1994).

8.2.4. Uneven Allocation of Wealth

In Gates's opinion (1999, p. 8), "two-tier societies and two-tier marketplaces are not the fertile soil in which robust democracies take root." Presently, 20% of this planet's population possesses the income of the remaining 80% (Leisinger and Schmitt, 2002). Impaired social unification coupled with the declining economic development is some of the problems being caused by the decrease in production and skills because of inequality. Discrimination in society leads to breakups in the family, felony, terrorism, substance abuse, reliance on the aid provided by the country, etc. Social inequalities are not only concerned with practical affairs but with aspects of modesty and virtuousness as well. The poorly concocted diplomatic international collaborations; the ever-rising economic domination of the World Trade Organization is the reason behind these growing disparities (UNDP, 1999). The entire economy is endangered as a large section of the community remains disconnected from the financial institutions and services. Extreme differences like these, between, and within the state have never been encountered previously (UNDP, 2000).

8.2.5. Effects on the Environment and Other Latest Challenges

Growing urban population consequences, climatic transitions, the emergence of unfamiliar diseases, such as the outbreak of AIDS, are some of the latest issues being dealt with globally. Controlling them requires significant change and creativity. Figure 8.2 shows the rise to be more than 6 billion to three folds, in the number of people living in cities.

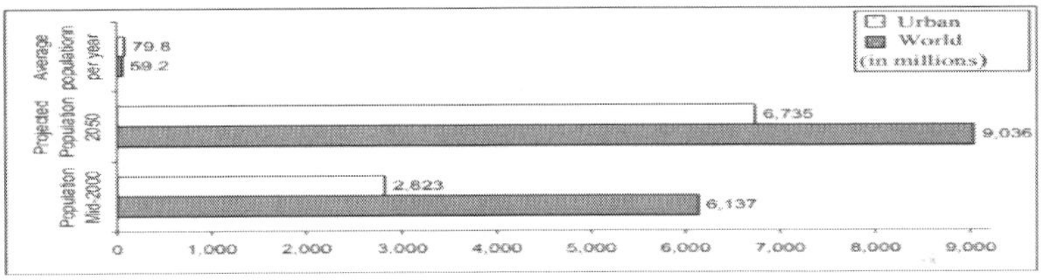

Figure 8.2: Trends of the world population.

Source: http://www.prb.org/Content/NavigationMenu/Other_reports/2000-2002/2001_World_Population_Data_Sheet.htm#highlights.

Even though the discussed transitions have existed for more than half a century, humanity is now confronted with even greater and severe social and ecological constraints (Roome, 1998). Therefore, ultimately, this man-made environment will have to face any future modifications produced by the increasing urban population and climate, with SD being the sole solution to all these problems (Enmarch-Williams, 1996; Blair, 2005).

8.3. THE ORIGIN OF SUSTAINABLE DEVELOPMENT (SD)

'Living in harmony with nature and within society'-The modern concept of sustainability happens to have stemmed from the substance and framework provided by numerous religious principles, classical thought, knowledge, and historical downfalls and therefore have contributed to shaping the idea of sustainability people hold today (Adetunji et al., 2003). Hence, the literature review points out that this concept has not been recently developed. The extensive transformation brought by the economic design introduced at the beginning of the 1970s has received greater attention (Meadows, 1972;

Schumacher, 1973), as compared to the one enforced in the 1960s because of the latter's negative economic influence (Carson, 1962; Erhlich, 1968).

The World Conservation Strategy, 1980 was endorsed by The International Union for Conservation of Nature (IUCN) in association with The United Nations Environment Program (UNEP) and the World Wildlife Fund for Nature to provide a unified approach towards conservation by merging the ecological and progressive affairs (Tryzna, 1995). Its prime focus was to strengthen the previously vague belief of SD and determine its central idea (Khosla, 1995). According to Holmberg (1994) sustainability as a concept could have been introduced by the IUCN. Yet, it was presented in the field of politics by The World Commission on Environment and Development (WCED) via the Brundtland Report (1987). Efforts made worldwide and by the government of the United Kingdom for the advancement of SD can be seen in Figure 8.3 (Geisinger, 1999; De Boef et al., 2010; Onoji et al., 2016).

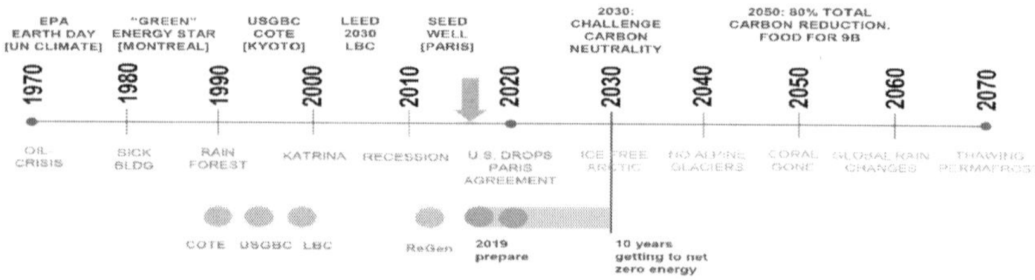

Figure 8.3: The term of sustainable development challenges.

Source: https://www.bnim.com/blog/bnim-s-2020-sustainability-action-plan.

An authentic examination was currently carried out to review the growth of SD in the UK since 1999. After careful consideration, the declaration was made; 'Shows Promise, But must try harder' (SDC, 2004). Despite sustainability's boosting political support during the past 40 years and the struggle to unite the ecological and communal challenges with economic progress, the achievement of the desired goals remains far out of reach as predicted by different studies. Meanwhile, international surveys concluded "little evidence of a massive shift in attitudes and actions on the part of all major players upon which the realization of a SD process depends. Individual, political, and entrepreneurial inertia as well as tactical behavior continue to delay a halt of environmental destruction and resource mismanagement" (NFSD, 2002, p. 2). Keeping this in view, the chapter reviews the reasons behind its unsatisfactory growth and their proposed solutions (Smith, 2006; Munyi and De Jonge, 2015).

8.4. HURDLES IN THE PATH OF SUSTAINABLE DEVELOPMENT (SD)

8.4.1. Documented Inequity Between the Two Fragments of the Concept

Environment and growth; two currently well-defined branches, provide the fundamental basis for sustainability's evolution. However, these branches were not distinguished up until the 1960s (Stevenson, 2006; Miles and Singal, 2010). Modification in the perception of accomplishments and issues of both the disciplines led to the awareness of the fact that they are both interconnected and influenced situations simultaneously. Hence, the reason behind sustainability's substandard implementation may be the difference in the origin as well as the philosophical interpretation of growth and environment (Lélé, 1991; Sharpley, 2000). For a better understanding of the demands of SD's effective enforcement the importance of acknowledging the differences in its approach is very crucial (Elliot et al., 1994).

8.4.1.1. Change in the Understanding of the Development

It was believed that issues of the South could be resolved with the help of technology, monetary support, and transference of skills after a favorable cause and effect relationship was implied between progress and economic advancement in the 1960s. Gross National Product (GNP) was used to assess the growth of the economy (Elliot et al., 1994; and Radclift, 1987). Some improvement was shown by GNP in the 1970s. In the 80s however, attention was also given to 'Controlling the population' and 'Equal distribution of wealth' due to issues such as discrimination, poverty, and the growing population. These problems created additional demands on the development of the economy. Managing population growth and equal allocation of wealth happen to be basic principles of any developmental program just as the phrases 'Growth with Equity' and 'Redistribution with Growth.' It was soon realized that all the divisions such as agriculture, industry, etc, called for investment to bring about progress and one model alone would not be able to make that happen 1980s led to comprehending development in a new light and being regarded as a multi-faceted belief encompassing extensive reformations in the community along with material welfare of the whole society. Apart from the social and economic aspects population, natural resources and their influence on the ecosystem also need to be included in development for it to be sustainable (Radclift, 1987).

8.4.1.2. The Environment's Modified Perceptions

As the numerous consequences of the industrial world's unsustainable system of manufacturing and utilization started to surface, relevant organizations broadcasted works related to preserving the environment in the 1960s (Carston, 1962; Erlich, 1968). Rising population, increasing pressure on the natural resources in the developing nations and the risks they pose on the world's environmental attributes further aggravated the situation. Unfortunately, the developed countries (DCs) did little to support the idea of environmental conservation due to the following facts (Radclift, 1987): Assets were considered limitless.

- Development was restricted due to doubts behind its proposal;
- Industrialization was the only path towards progress;
- Industrial countries were the cause of environmental issues therefore; they needed to deal with it;
- Environmental deterioration was an unpreventable outcome of industrialization.

8.4.1.3. The Advent of Development and Environment

To unify the urge for economic prosperity in the North and economic progress in the South with the environmental and social downfall is a serious political conflict of the Brundtland (Adetunji, 2003). As proven in numerous international environmental goals and political conferences, this disproportion in history has caused dissimilarities in principles, opinions, outlook, and understanding, judgment, and vision present today.

8.4.2. Lack of Clarity and Contradictions of the Concept

Various definitions of SD have been provided in this chapter because of the disagreements about its terminology. It can be described as an overlapping, multidisciplinary concept characterized by radical differences of conflicting targets (Dovers and Handmer, 1993). 'Sustainability' and 'development' as a concept are two opposing elements that usually lead to structural variation. Instead of putting together the aggregate of both these ideas, multi-faceted variables and sub-variables can be culminated by coupling these two words together (Samson, 1995).

Eight differences detected by Dovers and Handmer (1993) are:

- **Economic Growth versus Ecological Limits:** 'Sustainability' and 'development' are incompatible.

- **Individual versus Collective Interests:** The reunion of two different concepts. Safeguarded though the political system, the western ideology is characterized by exclusive independence. Whereas, sustainability is a communal issue brought about by the totality of individual choices made by men.
- **Cause versus Cure:** Technological and cultural conflict.
- **Intergenerational versus Intragenerational Equity:** Redistribution of resources is an ecologically defined but politically impossible trade-off. Organically explained but the politically unattainable transaction of the rearrangement of resources.
- **Humility versus Arrogance:** Doubtful decision making despite the growing knowledge about the world's environmental calamities.
- **Adaptability versus Resistance:** Reforms being met with various forms of persistence.
- **Optimization versus Spare Capacity:** Increased efficiency is anti-sustainability.
- **Diversity versus Purpose:** Possible clash between intentional action and the democratic range (Table 8.1).

Table 8.1: A Brief Outline of the Remarks on the Contorted Theories

Comments	Commentators
Vague concept and with a wider scope for clarification	Meadowcroft, 1999; Oldeman, 1995; O'Riordan Voisey, 1998; Hill and Bowen, 1997; Gatto, 1995
Unclear, confusing, and imprecise	Dovers and Handmer, 1993; Mullaney and Pinfield, 1996
A breeding place for arguments, motherhood, and apple pie	Brandon, 2000; Cecchini, 2000
Discordant and insufficient	Pearce, 1989; Daly, 1996
A reaction resulting from human activity focused on the betterment of mankind alone	Kirby, 1995

8.4.3. Time Needed for Cultural Reformation and the Absence of Political Will

Cultural change is essential as the achievement of sustainability will require an extensive and collective approach. Political and social bodies need to band together on a regional, national, and international level to accompany change needed for striving towards sustainability. The political drive to execute the numerous conditions that had been attained specifically during and after the Rio Submit has remained undetectable (Blair, 2005). Motivational conferences related to biological diversity, forests, and atmospheric conditions were held after the international agreement on ozone-depleting chemicals like CFCs and national contracts on sulfur dioxide generally showed hope that mankind was headed towards the right course. The result of these three problems, however, reveals a loss of enthusiasm (Dresner et al., 2002).

8.4.4. Distinct Perspectives and Clashes of Interest between Significant Players

The present-day formal consensus system and democratic decision making is being questioned due to its incapability of attaining sustainability. The main causes behind the unsatisfactory program are the clashes in interests and understanding disparities between the important figures on the environment's capability to withstand the effect of man's activities. The four corresponding approaches of the environment represent a combination of observed facts and materialism (Samson et al., 1995). They can be utilized to distinguish the theoretical discussion based on the impact of environmental transition and international progress as acquired from the works of Samson (1995), Thompson (1990), and Rayner (1991). The bowl presented in the picture shows the perfect standard placement, illustrating the four geomorphologies of the environmental perspective in Table 8.2 (Lorne, 2009; Hwang and Tan, 2012).

The differing views and conflicts of interest among major players on the ability of the environment to endure human impact are the major reasons for the inadequate policy. This has raised concerns on the effectiveness of the current democratic decision-making and consensus-seeking mechanism to the achievement of SD. Drawing from the work of Thompson (1990), Rayner (1991), and Samson (1995), the conceptual debates on the effect of the global development and environmental change can be characterized using the analogy of four competing views of environment, which represent a mixture of physical and perceived reality (Samson et al., 1995). As depicted in Table 8.2, the illustration comprises four topologies of environmental views, demonstrated by a bowl situated in a landscape to represent ideal-typical positions (Lorne, 2009; Hwang and Tan, 2012).

Table 8.2: A Brief Overview of the Four Competing Points f View

Views of Nature	**Visualization**	**Comment**	**Views**	**Institution/ Shared**	**Solution**	**Example (Tolba et al., 1992)**
I. Environment is robust		The ball will always come back at the base of the bowl to maintain the balance, no matter what human do to the landscape as the environment is strong and flexible (Simon, 1981).	Individualist: Fresh and multiple possibilities for human insight can be provided from this positive challenge from environmental issues.	Profession and economy	Sustainable technology and the free market. The market should be driven on the proper route by the invisible hand. The Govt. should have a neutral perspective.	Restoration of the initial thick lust forests in Central America due to overuse by the primitive communities.
II. Environment is fragile	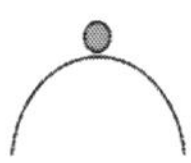	A slight knock to the environment can send it crumbling, since it is very frail and harsh (Goldsmith, 1993; Devall and Sessions, 1985).	Egalitarian: Many adverse effects of human activities can be seen through climatic changes which can ultimately cause permanent destruction Earth	Deep ecologist	Existing in peace with the environment, replacement of anthropocentric systems with biocentric equity,	The issue of urban growth and desertification
III. The environment is robust within limits		The ball can tip over the edge if the environment's definite limits are surpassed. Within those limits it is forbearing and resolute (Brundtland Report and National/ local strategies).	Hierarchies: Careful scientific perception is required to manage natural resource and environmental deterioration for the prevention of a worldwide disaster	Government/ Political organization	Incorporation of externalities; principles and administrative processes; scientific investigation	Over hundreds of years a large North Nigerian city, Kano has experienced occasional droughts. Important nutrients promoting faster renewal are liberated by forest fires.

IV. Environment is chaotic		A ball on an even plane. The environment disregards any kind of regulation, authority, or mitigation. It is uncontrollable, unreliable, and unsystematic.	Fatalist: The environment is chance influenced cornucopia with sheer luck	Benefactors mostly do not state their ideas	The best answer is to take no action	

The Europeans' debate is based on the concept 'environment is robust within limits' whereas the Americans argue that 'environment is robust.' Interpretation of an idea is based on an individual's unique character to defend their selected course of action, surroundings, and all contradictory outlooks are rogue figures, planned, and misplaced attempts to hinder economic progress. This brings to mind, the disappointing development after the Rio summit is credited to the split on the topic of climatic change and forests between America-Europe and South-North (Huesemann, 2003; Zelenika and Pearce, 2014). The distinct individualized reasoning presented by the egalitarian (deep ecologist), hierarchies (political institutions), and individualist (business and industry) validates their activities. As shown in Table 8.2 whenever the multisided and unexpected parts of nature become active, curiosity grows as normally it fits every one of these concepts. Hence, confirming the validity of these numerous views alternatively (Tang et al., 2011).

8.5. POTENTIAL SOLUTIONS

There are differences between the processes of consensus-seeking, long-term planning and democratic decision-making associated with SD. Coupling progress with several perspectives of environmental proficiency through prevailing 'political' and 'market' forums have proven to be limited in range, as implied by the analysis up till now (Schomberg et al., 2002). Agreements related to climatic change are being established mostly on political achievability rather than on the permissibility of the environment by many national delegates. The causes behind that are: (i) the rise in the probability of being re-elected in the national government particularly when the economic goals are linked with the environmental ones. There are strong grounds on which many political leaders can deny or accommodate the smallest of environmental operations that leave minimal impacts on the economy and still be exhibited as a communal achievement. Some examples are the unwillingness to minimize CO_2 emissions and the refusal of the Kyoto agreement by the Bush-administration. (ii) Increasing accumulation of authority in the hands of a small number of multi-national companies through globalization (Dresner, 2002).

A need to shelter the restrictions of the authoritative political systems and markets and clashing value has been pointed out in the chapter and by numerous authors (Birkeland, 1996; Keijzers, 2002; Schomberg, 2002). This can be done formulating calculated processes and international standards of harmonized commitment by supporting the need for the alleged third ground of the societal conflict resolution. Contradictory to the current system of compromise and vote, the third arena is established on the management of change, decision-making mechanism centered on discussion and intervention. The third arena is built on the features mentioned below (Schomberg et al., 2002):

- Constitutional change or modification in the political framework to house procedures for technology examination, the latest types of social debates, and the progress of transpersonal science.
- To establish a link between the prevailing social debate, political decision-making mechanism, and a certain subsystem, the need for a program for technological examination to supplement them should be recognized.
- Increase in public debates on all stages to offer a connection between system and subsystems-political decision-making process, law, science, and politics.
- Science for sustainability is an intricate institution, enveloped by scientific obliviousness and doubt. To guarantee a superior policy system, sustainability, administration, and safety regulation should be interconnected. There is an argument for a new kind of instrument to encourage continuous communication for science policy. To evaluate the nature of information instead of truth enclosed in every scientific declaration, a new decision-making instrument for policymakers is also required.
- A normative (deliberative) plan founded on the understanding of recognized normative aims, as the purpose of withdrawal can facilitate searching for a way between policy and scientific level. Planned processes like the implementation of safety regulations can assist in leading to an agreement on such aims.

- System innovation offers a way of reaping the advantages of sustainability. There is a need of 'transition management' with the main components of construction of transitional aims and the utilization of system management focused on the fundamentals of doing by learning and learning by doing, to solve the momentary motive of public and private branches and aim of sustainability in the long run.
- Visionary enterprises: the future organizations are the ones that begin to incorporate long term planning in their business as the number of green consumers keep on increasing (Table 8.3).

Table 8.3: Aspects of Societal Conflict Resolution

Dimensions	Individuals	Individuals/Society	Individuals/Society
Interests	Priorities/Wants	Rights	Needs/Responsibility
Standards	Effectiveness	Equality/Social Justice	Ethics of Responsibility- Safety regulations/ Search for sustainability
Processes	Trade	Vote/Consultation	Discussion/arbitrate/managing change
Arena	Market	Politics/ compliance of stakeholder	Long term Planning related to sustainable development/ Global negotiations

Source: Schomberg (2002).

8.6. CREATIVE SOLUTIONS FOR SUSTAINABILITY

Purposeful inventive activities and uncommon tactics used to acknowledge the ecological, communal, and economic requirements comprehensively to motivate change. The thought behind these solutions represents an ideal change that would result in quicker socioeconomic modifications by encouraging nations to take action. Generally, these developments concentrate on technological advancements however; innovative solutions are aided by non-technological reformation too. Thus, it must not be kept in mind that innovative solutions are limited to inventions or confined to technology (Lee, 2001; Ashford and Hall, 2011).

These solutions integrate revolutionary ideas all overvalue chains in several divisions. They are transformed for the environmental issue and made feasible for products, processes, ecological assistance, organizational framework, and market strategies which would increase efficiency, ecological benefits (ecological and economics), and sustainability. They integrate all the components of sustainability: environmental, economic, social, and nurturing alliances all over the value chain while also emphasizing lifestyle thinking (Lorne, 2009).

By the attainment of knowledge, skills, interchange of information, and contribution from a partnership with value chains innovative solutions boost technical capacity. Decisions related to investments are largely being based on sustainability. These solutions are profitable and therefore are drawing new and old investments as they happen to be flexible and practical in the long run. They are meeting those demands of the market that had not been fulfilled before (Quist and Vergragt, 2004).

Innovative solutions can bring about economic prosperity and increased employment opportunities by recognizing the possibilities for growth through sensible production and consumption, reducing wastes, water, power, and material proficiency, profiting through value chains, hence boosting productiveness and productivity. A well-designed international policy is required to steer innovative solutions to provide it a stimulating market and political atmosphere without any boundaries to hinder its progress.

8.6.1. Significance of Innovative Solutions to Attain SD

It is now believed internationally that the execution of sustainable development goals (SDGs) calls for a unified approach. However, how it is supposed to be applied practically is a challenge. Innovation has been recognized as one of the fundamental motivators of global economic progress, employment generation, and productivity by Agenda 2030. All important organizations have been urged to engage in resolving the issues of SD through invention and creativity. A great degree of economic efficiency can be achieved (SDG 8) through policies targeted on development facilitating innovation.

Energy-based innovations such as using clean energy (SDG 7) can lead to decreased emission levels hence aiding in the achievement of Developmental Goal 13. Innovative solutions promote food security (goals 1, 2, 3) and climatic adaptation (SDG 13 and 15) by unifying ecosystem-based adaptation. By improving environment-friendly and other wide-ranging jobs, it also brings about employment and economic progress (SDG 8) (Dincer, 2000; Sahlberg and Oldroyd, 2010).

SDG 9 can be pursuit by funding cleaner production and resource-efficient technology, along with preservation of natural resources (e.g., energy (SDG 7)) and water (SDG 6); land and marine ecosystems (SDG 15 and

14), decreases utilization of chemicals to manage environmental pollution, reliable patterns of consumption and production, better occupational health and precaution (SDG 3) by more effective environmental conformity, and producing more jobs (SDG 8). Growth in economic prosperity, security, and decline in shortage of resources (Goal 16) are some of the outcomes if the previously mentioned solutions are practically applied thus improving inclusive nations and harmony (Bouma, 2014; Lukman et al., 2016).

To uphold innovations, Goal 9 is the basic aim which includes improvement in technological abilities, supporting the development and significantly expanding the amount of development and research employees per every 1 million people.

8.6.2. Focus Areas Requiring Innovative Solutions

Interconnecting partnerships and sustainability along with modification in life-cycle thinking are needed for managing environmental issues in an innovative style (Bratt et al., 2013; Czarnecki, 2017). Innovative solutions have been endorsed by the UN Environment Programme recently, as an answer to global environmental problems. Some of the ecological issues that need to be dealt with the help of innovative solutions have been mentioned below (Lorne, 2009; Witkowski, 2017):

- Food insecurity;
- Sand and dust storms;
- Anti-microbial resistance;
- Pollution, including indoor air pollution;
- Lack of biological variety;
- Energy insecurity;
- Unfeasible patterns of consumption and production;
- Environmental change and deterioration;
- Waste.

8.7. AREAS OF INNOVATIVE SOLUTIONS REQUIRING FOCUS

Policy and non-policy interventions to create an enabling environment for innovative solutions to environmental challenges and SCP are needed. Having them in place stimulates investment, financing, and enables the transition to SD. Implementation of the interventions also calls for change in governance frameworks: policies, laws, and institutions to trigger change, on a foundation of equity, inclusiveness, and participation. Some of the interventions are (Boer, 1995; Subasinghe et al., 2009):

8.7.1. Economy

To build an environment that facilitates innovative solutions for ecological challenges and SCP there is a requirement for policy and non-policy negotiations. Their establishment boosts finance, funds, and allows movement towards SD. Their application demands transformation in administrative structures: organizations, policies, and rules and to initiate change, based on equality, collaboration, and comprehensiveness. A few of the interventions are explained in following sub-sections (Boer, 1995; Subasinghe et al., 2009).

8.7.1.1. Circular Economy

Economic approaches are transitioning to models of to circular economy from current linear ideals. It emphasizes on increasing the lifespan of products, reforming resources (utilization of harmless, reusable, renewable), treating waste to gain resources, future planning, reevaluating business designs; all of which require innovative solutions.

8.7.1.2. Sharing Economy

Access is the new ownership-a growing number of individuals recently believe that at times, renting provides more benefits than owning. Nowadays people's mindset has started moving on from the idea that a person's reputation in society is eminent by the amount of goods they own. For example, collective pest management facilities are being provided to farmers using unprocessed products and material by an organization, so the firm delivers the service to its customers except for employing the principle method for selling pesticides. The present transition towards collective utilization and a sharing economy is a representation of an innovative solution (Tangcharoensathien et al., 2015).

8.7.2. Adjustment of Policy

A setting that encourages innovative solutions can only be made if policies though out numerous fields are harmonized. For instance, to make sure, that

EBA procedures such as agroforestry are fused into the standard policies of agriculture (SDGs 2, 13, 15), the agricultural regulatory commission will have to work alongside the forestry ministries. Agriculture policies will have to coordinate with the policies of land, power, industry as well as private investors to guarantee transverse strategies that stimulate financers through each of the two, state, and non-state actors in factories and clean processing industries to bring funds in clean energy-driven value addition (Azmat, 2013).

8.7.3. Innovative Financing Schemes

The Addis Ababa Action Agenda strives to upgrade social innovation (accompanied by innovative approaches of primitive peoples) to facilitate sustainable employment and social welfare. It invites organizations to use their innovation and creativity to resolve the issues of SD.

Latest financial streams (green bonds; eco-friendly energy systems; mobile banking, green mortgages; retro-fittings in buildings) are required.

8.7.4. Innovative Partnerships

The world's most essential element in terms of wealth is human capital. Innovative solutions demand an all-encompassing alliance, where individual, non-state, state institutional shareholders attracted from interrelated fields collectively have an advantageous cooperation to steer the higher promoters. Goal 17 suggests these collaborations, which are proving functional as described in the above-mentioned examples.

8.7.5. Implementing Synergy

When collaborations, strategies, markets, and financial innovative solutions are utilized together to take steps to analyze communal, economic, and environmental issues, then synergistic implementation is taking place. For instance, combining characteristics of preservation and processing to prevent food loss and wastage results in the improved generation of jobs and food security. Climatic targets can be attained by using renewable energy (RE) to fuel these procedures. We then collectively have communal, ecological, and economic targets being fulfilled simultaneously. Synchronically, though improved food security, added production, and greater material proficiency SCP aims can be attained. Integrated efforts are necessary on both national and international levels for the application of this concept, primarily through the management of the available systems that supplement innovative solutions for SCP and environmental issues. Some of the significant ones among these mechanisms are:

- To research for additional innovative solutions for environmental issues (both current and upcoming) the 10-YFP Secretariat, through its Trust Fund, is increasing its contribution to SCP (which is hosted by the UN Environment Programme). This indicates that the present system of allotting resources in the Trust Fund needs to take into perspective, the demands of the community to accommodate and accept the innovative solutions so that they are integrated into developmental designs (including national SCP plans, where they exist).
- Facilitating Research and Development, particularly in transforming, developing economies, and nations that concentrate on innovative solutions according to the national demands. Proper application requires imitation of some pre-existing designs, concepts, and systems that have been exercised or exist somewhere. It is essential to share the developments, results of research and information that can improve innovative solutions for ecological problems. Evenly affecting developing and developed states, the forthcoming environmental issues are transborder.
- Framework of innovation that encourages the utilization of lesser resources (resource efficiency) is very significant-Developing nations can benefit from the interchange of understanding and experience related to the resolute and environment-friendly structure (e.g., design of green buildings) so that they can enforce the needed innovative solutions for this cause.
- Taking initiative, developing countries have set up innovative financing techniques (e.g., mobile money) that need to be improved to cause a decline in ecological problems that still plague this class of nations and promote the acquisition of a few SDGs. Aiming at these promotions is crucial for DCs for producing synergies that have a deeper and wider impact.

8.8. ROLE OF THE UNITED NATIONS

"Achieve international co-operation in solving international problems of an economic, social, cultural, or humanitarian character and in promoting and encouraging respect for human rights and fundamental freedoms for all without distinction as to race, sex, language, or religion." were the primary preferences of the UN from the beginning of 1945. Improving human welfare is still one of the most important objectives of the UN. SD -progress that contributes to communal interest, success, facilitates the economy, and environmental preservation-has been recognized by several nations as the key to moving ahead for uplifting the lives of people all around the world after the modification in the concept of development (Williamson et al., 1999) (Figure 8.4).

Figure 8.4: A tree being planted in 2012 by Grace, a chief member of the forest community in Ainabkoi, Kenya, and a farmer from the village of Kipilat.

Source: https://www.un.org/en/sections/what-we-do/promote-sustainable-development/.

Just twenty years ago, the people residing in developing countries were living in severe poverty. However, much as changed since then. UN's millennium development goals (MDGs) have considerably led the way to advancement by decreasing poverty to half of what it was before. Hence, a 2030 Agenda for SD has been approved, to fulfill the demand of eliminating poverty. Meanwhile, a well planned, international climatic contract promoted by the UN in 2015 is being enforced as climate change presenting emerging challenges to the world's developmental goal. For sufficient supplementation of climatic activity and SD targets, the UN is planning on formulating a fund for a progressive framework (Mochizuki and Fadeeva, 2008).

8.8.1. A Program for Sustainable Development (SD)

2015 Time for Global Action for People and Planet, a portal for SD was set in motion by the UN in 2015. It focuses on the UN's SD agenda after 2015 and it contains details of the actions taken by the UN to combat climatic change and other similar problems. The Secretary-General's report that represents the insight for the post-2015 agenda and demonstrates ways, in which critical issues can be highlighted, has been featured on the website. UN's new post-2015 SD agenda began at the SD Summit in September 2015. Only a minority was relieved from poverty with the help of MDGs. The endeavor that began with the MDGs will be brought to completion by the UN's new SDGs (Linnér and Selin, 2013).

8.8.2. SDG Summit 2019

At the United Nations Headquarters in New York, a meeting will be held on 24 and 25 of September 2019. Heads of State and Government will look into and deeply exam the success in the execution of 17 SDGs and the 2030 Agenda for SD. From the time 2030 Agenda was implemented in September 2015, this is the first gathering of the UN summit on the SDGs.

8.8.3. The Millennium Development Goals (MDGs)

Global leaders devoted their countries to attaining the eight MDGs by 2015, in September of 2000. These goals include stopping HIV/AIDS from spreading, imparting primary education universally, and reducing severe poverty by 50%. Different objectives have been initiated by the Secretary-General which involves Every Woman, Every Child, and Zero Hunger Challenge, to ensure their attainment. Even though achieving most of these targets has been majorly successful, more is needed to be done.

8.8.4. Change in the Climate

A warning has been issued by the UN's Intergovernmental Panel on Climate Change (IPCC) related to the risks of climatic change and they also mentioned the importance of finding the right course of action before it is too late. With a significant increase in the emission of greenhouse gases, the planet is moving towards the point where the standard global temperature in this century will be raised to more than 3°C. Melting glaciers, intense weather conditions, and increasing sea levels are some of the effects of climatic change that are being felt globally. Sustainability is crucial for nullifying the effects of these changes. UN is working to support dialogue under the UN Framework Convention for a climatic settlement, analyzing climatic science and aiding communities and countries in creating an adaptable climate and decrease emissions. Secretary-General Ban Ki-Moon has started several programs. Out of those programs, one is called Sustainable Energy for all, which includes enhancing power efficiency, greater use of sustainable means of energy, and assist people in obtaining clean energy. A contract based on attaining new global climatic change has been encouraged by the UNFCCC Secretariat as it offers a way forward by limiting the increase in temperature to below 2° or even 1.5°. Taking place in New York in September 2014, The Climate Summit assisted in increasing awareness about the significance of climatic transition by rallying support for a contract and brings about the activity before the meeting in Paris, 2015.

8.8.5. Disaster Risk Reduction

Strength should be the core of sustainability as catastrophes can tear down societies in the blink of an eye. The United Nations Office for Disaster Risk Reduction (UNISDR) collaborates with authorities and other shareholders to guarantee a decrease in lives lost due to catastrophes and resources of nations and societies. Near the point where the disastrous tsunami and earthquake in Japan took place, a gathering proceeded with the intent to decrease the threats from disasters by embracing the Sendai Framework for Disaster Risk Reduction 2015–2030. Visiting the portal can add to the existing knowledge of disaster risk reduction.

8.8.6. Inspiring Women and Gender Equality

Girls and women still suffer from the hands of abuse and prejudice in all parts of the world even after the progress towards equality among genders through the UN's MDGs. Gender equality, may not be a basic human right but it forms the basis of a harmonious, sustainable, and successful planet. The purpose of UN Women is to motivate women, obtain equality between both the genders as partners, and eradicate discrimination against girls and women and receivers of development, harmony, protection, charitable deeds, and human rights.

8.8.7. Major Offices of the UN and Their Plans on Working for the Development

Working with shareholders and authorities The UN Department of Economic and Social Affairs assists nations around the globe to fulfill their communal, ecological, and economic aims. Continuous efforts of the United Nations Development Programme with individuals belonging to all levels of the community aided in building nations that can endure calamities and steer and maintain the progress that enhanced the standard of life for all. World Health Organization (WHO), UNICEF, UN Environment Programme, UNESCO, and Food, and Agriculture Organization are some of the firms of the UN that function for specific types of development.

8.8.8. What UN Bodies Work to Improve the Status of Development?

The General Assembly's Second Committee (Economic and Financial) handles matters related to human settlements, economic prosperity, globalization, information, and communication technologies, and elimination of poverty.

REFERENCES

1. Adetunji, I., Price, A., Fleming, P., & Kemp, P., (2003). The application of systems thinking to the concept of sustainability. In: *The Proceeding of the Association of Researchers in Construction Management (ARCOM)* (Vol. 1, pp. 161–170), University of Brighton, UK.
2. Annan, K., (2001). *United Nations, Economic and Social Council, Commission on Sustainable Development Acting as the Preparatory Committee for the World Summit on Sustainable Development, 1*, 1–30. www.cop4.org/resource/docs/cop3/l07a01.pdf (accessed on 1 May 2020).
3. Ashford, N. A., & Hall, R. P., (2011). The importance of regulation-induced innovation for sustainable development. *Sustainability*, *3*(1), 270–292.
4. Azmat, F., (2013). Sustainable development in developing countries: The role of social entrepreneurs. *International Journal of Public Administration*, *36*(5), 293–304.
5. Birkland, J., (1996). *Ecological Government, in Technology and Society Magazine, 2*, 21–28.
6. Blair, T., (2005). *Forward, in DETR, Securing the Future, Delivering UK Sustainable Development Strategy* (Vol. 1, pp. 1–44). The Stationery Office, London.
7. Boer, B., (1995). Institutionalizing ecologically sustainable development: The roles of national, state, and local governments in translating grand strategy into action. *Willamette L. Rev.*, *31*, 307.
8. Bouma, J., (2014). Soil science contributions towards sustainable development goals and their implementation: Linking soil functions with ecosystem services. *Journal of Plant Nutrition and Soil Science*, *177*(2), 111–120.
9. Brandon, P. S., (2000). Sustainability in management and organization, the key issues? In: *Proceedings, Conference on Cities and Sustainability, Sustaining our Cultural Heritage* (Vol. 1, pp. 11–16). Kandalama, Sri Lanka.
10. Bratt, C., Hallstedt, S., Robèrt, K. H., Broman, G., & Oldmark, J., (2013). Assessment of criteria development for public procurement from a strategic sustainability perspective. *Journal of Cleaner Production*, *52*, 309–316.
11. Brundtland, G., H., (1987). *Our Common Future, Report of the World*

Commission on Environment and Development (Vol. 1, pp. 1–40). New York.

12. Carson, R., (1962). *Silent Spring* (Vol. 1, pp. 1–30). Hamish Hamilton Publishers, London.
13. Cecchini, A. B., (2000). On certain linguistic imperfections and the desire to avoid immortality, notes on the concept of the sustainable city. In: *Proceedings, Conference on Cities and Sustainability* (Vol. 1, pp. 71–78). Sustaining our Cultural Heritage, Kandalama, Sri Lanka.
14. Club of Rome, (1991). *The First Global Revolution* (Vol. 1, pp. 1–14). The world twenty years after, the limits to growth, by King A., Schneider B., Simon & Schuster, New York.
15. Czarnecki, L., (2017). Innovation in construction materials engineering versus sustainable development. *Bulletin of the Polish Academy of Sciences Technical Sciences*, *65*(6), 765–771.
16. Daniels, P. L., (1992). Barriers to sustainable development in natural resource-based economies: Australia as a case study. *Society and Natural Resources*, *5*(3), 247–262.
17. De Boef, W. S., Dempewolf, H., Byakweli, J. M., & Engels, J. M. M., (2010). Integrating genetic resource conservation and sustainable development into strategies to increase the robustness of seed systems. *Journal of Sustainable Agriculture*, *34*(5), 504–531.
18. Devall, B., & Sessions, G., (1985). *Deep Ecology, Living as if Nature Mattered* (Vol. 1, pp. 1–33). UT, Peregrine Smith, Salt Lake City.
19. Dimbleby, J., (2000). *From Here to Sustainability* (Vol. 1, pp. 1–23). Politics in the Real World, London.
20. Dincer, I., (2000). Renewable energy and sustainable development: A crucial review. *Renewable and Sustainable Energy Reviews*, *4*(2), 157–175.
21. Dovers, S. R., & Handmer, J. H., (1993). *Contradictions in Sustainability, Environmental Conservation, 20*(3), 217–222.
22. Dresner, S., (2002). *The Principle of Sustainability* (Vol. 1, pp. 1–44). The Earthscan Publications, London.
23. DTI, (2004). *Sustainable Construction Brief 2* (Vol. 1, pp. 1–43). http://www.dti.gov.uk/construction/sustain/fb.pdf (accessed on 1 May 2020)
24. Dyllick, T., & Hockerts, K., (2002). Beyond the business case for corporate sustainability. *Business Strategy and the Environment, 11*,

130–141.

25. Elliot, J. A., (1994). *An Introduction to Sustainable Development* (Vol. 1, pp. 1–23). The Developing World, Routledge, London.
26. Enmarch-Williams, H., (1996). *Environmental Risks and Rewards for Business* (Vol. 1, pp. 1–43). Chichester.
27. Erlich, P., (1968). *The Population Bomb: Friends of the Earth Book* (Vol. 1, pp. 1–29). London.
28. Fussler, C., & James, P., (1996). *Driving Eco-Innovation: A Breakthrough Disciple for Innovation and Sustainability* (Vol. 1, pp. 1–33). Pitman, London.
29. Gates, J., (1999). *The Ownership Solution* (Vol. 1, pp. 1–44). Penguin, London.
30. Gatto, M., (1995). Sustainability, is it a well defined concept? *Ecological Implications, 5*(4), 1181–84.
31. Geisinger, A., (1999). Sustainable development and the domination of nature: Spreading the seed of the western ideology of nature. *BC Envtl. Aff. L. Rev., 27*, 43.
32. Goldsmith, E., (1993). *The Way: An Ecological World-View* (Vol. 1, pp. 1–33). Boston, Shambhala.
33. Gun, J. M., & Keller, W., (1990). Biological recovery of an acid lake after reduction in industrial emissions of sulphur. *Nature, 345*, 431–33
34. Hill, R. C., & Bowen, P. A., (1997). Sustainable construction, principles, and a framework for attainment. *Construction Management and Economics, 15*(3), 223–39.
35. Holmberg, J., (1994). *Policies for a Small Planet* (Vol. 1, pp. 1–43). Earthscan Publications, London.
36. Huesemann, M. H., (2003). The limits of technological solutions to sustainable development. *Clean Technologies and Environmental Policy*, *5*(1), 21–34.
37. Hwang, B. G., & Tan, J. S., (2012). Green building project management: Obstacles and solutions for sustainable development. *Sustainable Development*, *20*(5), 335–349.
38. IPCC, (1996), Climate change (1995). In: Watson, R. T., Zinyowera, M. C., & Moss, R. H., (eds.), *Impacts, Adaptations, and Mitigation of Climate Change, Scientific Technical Analysis, Contribution of Working Group II to the Second Assessment Report of the Intergovernmental*

Panel on Climate Change (Vol. 1, pp. 1–55). Cambridge.

39. IUCN/UNEP/WWF, (1991). *Caring for the Earth: A Strategy for Sustainable Living* (Vol. 1, pp. 1–44). Gland, Switzerland.
40. Jacobs, M., (1991). *The Green Economy, Environment, Sustainable Development, and the Politics of the Future* (Vol. 1, pp. 1–55). Pluto Press, London.
41. Kennard, R. M., (1995). Environmental challenge for engineers, Hong Kong perspective. *J. Profl. Issues in Engrg. Educ. And Pract., ASCE, 121*(2), 140–142.
42. Khosla, A., (1995). Foreword. In: Tryzna, T. C., (ed.), *A Sustainable World* (Vol. 1, pp. 1–39). Sacramento, IUCN.
43. Kirby, J., O'Keefe, P., & Timberlake, L., (1995). *Sustainable Development* (Vol. 1, pp. 1–55). The Earthscan Reader, Earthscan Publications, London.
44. Kühtz, S., (2007). Adoption of sustainable development schemes and behaviors in Italy: Barriers and solutions-what can educators do? *International Journal of Sustainability in Higher Education, 8*(2), 155–169.
45. Lee, K. F., (2001). Sustainable tourism destinations: The importance of cleaner production. *Journal of Cleaner Production, 9*(4), 313–323.
46. Leisinger, K. M., Schmitt, K. M., & Pandya-Lorch, R., (2002). Six Billion and Counting, Population and Food Security in the 21st Century (Vol. 1, pp. 1–36). International Food Policy Research Institute (IFRRI) and Johns Hopkins University Press, Washinton, D.C.
47. Lélé, S. M., (1991). Sustainable development: A critical review. *World Development, 19*(6), 607–621.
48. Linnér, B. O., & Selin, H., (2013). The United Nations Conference on sustainable development: Forty years in the making. *Environment and Planning C: Government and Policy, 31*(6), 971–987.
49. Lorne, F. T., (2009). Macro-entrepreneurship and sustainable development: The need for innovative solutions for promoting win-win interactions. *Environmental Economics and Policy Studies, 10*(2–4), 69–85.
50. Lukman, R. K., Glavič, P., Carpenter, A., & Virtič, P., (2016). Sustainable consumption and production-Research, experience, and development: The Europe we want. *Journal of Cleaner Production, 138*, 139–147.

51. MacRae, R. J., Hill, S. B., Henning, J., & Mehuys, G. R., (1989). Agricultural science and sustainable agriculture: A review of the existing scientific barriers to sustainable food production and potential solutions. *Biological Agriculture and Horticulture*, *6*(3), 173–219.
52. Malthus, T., (1798). *An Essay on the Principle of Population and Summary View of the Principle of Population* (Vol. 1, pp. 1–33). Penguin Books Ltd, Middlesex.
53. Meadowcroft, J., (1999). Planning for sustainable development, what can be learned from the critics? In: Kenny, M., & Meadowcroft, J., (eds.), *Planning Sustainability London* (Vol. 1, pp. 1–43), Routledge.
54. Meadows, D. H., Meadows, D. L., Randers, J., & Behrens, III. W. W., (1972). *The Limits to Growth* (Vol. 1, pp. 1–53), Universe Books, New York.
55. Miles, S., & Singal, N., (2010). The education for all and inclusive education debate: Conflict, contradiction or opportunity?. *International Journal of Inclusive Education*, *14*(1), 1–15.
56. Mill, J. S., (1848). *The Principles of Political Economy* (Vol. 1, pp. 1–23). Penguin, Harmondsworth, 1985.
57. Mochizuki, Y., & Fadeeva, Z., (2008). Regional centers of expertise on education for sustainable development (RCEs): An overview. *International Journal of Sustainability in Higher Education*, *9*(4), 369–381.
58. Muir, J., (1894). *The Mountains of California* (Vol. 1, pp. 1–43). Penguin, New York.
59. Mullaney, A., & Pinfield, G., (1996). No indication of quality or equity. *Town and Country Planning, 1*, 132–136.
60. Munyi, P., & De Jonge, B., (2015). Seed systems support in Kenya: Consideration for an integrated seed sector development approach. *Journal of Sustainable Development*, *8*(2), 161.
61. NFSD, (2002). *Sustainable Development: A Common Challenge for North and South* (Vol. 1, pp. 1–43). Novartis Foundation for Sustainable Development (NFSD). http://www.foundation.novartis.com/sustainable_development_print.htm#outlook (accessed on 1 May 2020).
62. Nordic, (2002). *Business Models for Sustainability, Experience, and Opportunities from the Nordic Partnership* (Vol. 1, pp. 1–53). WWF and Mandag Morgen, Copenhagen.

63. O'Riordan, T. R., & Voisey, H., (1998). *The Transition to Sustainability* (Vol. 1, pp. 1–43). The Politics of Agenda 21 in Europe, Earthscan, London.

64. Oldeman, R. A. A., (1995). Sustainable development is fuzzy development. *Nature and Resources, 31*(3), 1–15.

65. Onoji, S. E., Iyuke, S. E., Igbafe, A. I., & Nkazi, D. B., (2016). Rubber seed oil: A potential renewable source of biodiesel for sustainable development in sub-Saharan Africa. *Energy Conversion and Management, 110*, 125–134.

66. Pearce, D., Markandya, A., & Babier, E., (1989). *Blueprint for a Green Economy* (Vol. 1, pp. 1–23). Earthscan Publications, London.

67. Quist, J., & Vergragt, P. J., (2004). Backcasting for industrial transformations and system innovations towards sustainability: Relevance for governance. In: *Governance for Industrial Transformation. Proceedings of the 2003 Berlin Conference on the Human Dimensions of Global Environmental Change. Berlin: Environmental Policy Research Centre* (Vol. 1, pp. 409–437).

68. Radclift, M., (1987). *Sustainable Development, Exploring the Contradictions* (Vol. 1, pp. 1–36). Methuen, London.

69. Rayner, S., (1991). A cultural perspective on the structure and implementation of global environmental agreements. *Evaluation Review, 15*(1), 120–47.

70. Roome, J. N., (1998). *Sustainable Strategies for Industry* (Vol. 1, pp. 1–33). The Future of Corporate Practice, Washington, D.C.

71. Sahlberg, P., & Oldroyd, D., (2010). Pedagogy for economic competitiveness and sustainable development. *European Journal of Education, 45*(2), 280–299.

72. Samson, P., (1995). *The Concept of Sustainable Development* (Vol. 1, pp. 1–43). Discussion Papers, Green Gross International. http://www.gci.ch/DigitalForum/digiforum/discussionspapers/concept.html.

73. Schmidt-Bleek, F., (1994). *Wieviel Umwelt braucht der Mensch?* MIPS. Das Maß für ökologisches Wirtschaften, Basel. (in English as: *The Fossil Makers-Factor 10 and More* (Vol. 1, pp. 1–37). New York).

74. Schumacher, E. F., (1973). *Small is Beautiful* (Vol. 1, pp. 1–34). Blond & Briggs, London.

75. SDC, (2004). *Shows Promise But Must Try Harder, An Assessment by the Sustainable Development Commission of the Government's*

Reported Progress on Sustainable Development over the Past Five Years (Vol. 1, pp. 1–39). http://www.sd-commission.org.uk/news/resource_download.php?attach_id=SMDOB797JQMZFE-1OHI1PA-UFRDL3I (accessed on 1 May 2020).

76. Sharpley, R., (2000). Tourism and sustainable development: Exploring the theoretical divide. *Journal of Sustainable Tourism*, *8*(1), 1–19.
77. Simon, J., (1981). *The Ultimate Resource* (Vol. 1, pp. 1–36). Princeton, N. J., Princeton University Press.
78. Smith, A., (2006). Green niches in sustainable development: The case of organic food in the United Kingdom. *Environment and Planning C: Government and Policy*, *24*(3), 439–458.
79. Stevenson, R. B., (2006). Tensions and transitions in policy discourse: Recontextualizing a decontextualized EE/ESD debate. *Environmental Education Research*, *12*(3/4), 277–290.
80. Subasinghe, R., Soto, D., & Jia, J., (2009). Global aquaculture and its role in sustainable development. *Reviews in Aquaculture*, *1*(1), 2–9.
81. Tang, T., Karhu, K., & Hamalainen, M., (2011). Community innovation in sustainable development: A cross case study. *World Academy of Science, Engineering and Technology*, *73*(1), 396–403.
82. Tangcharoensathien, V., Mills, A., & Palu, T., (2015). Accelerating health equity: The key role of universal health coverage in the sustainable development goals. *BMC Medicine*, *13*(1), 101.
83. The Global, (2000). *Report to the President: Entering the Twenty-First Century* (Vol. 1, pp. 1–35). A Report by the Council on Environmental Quality and the Department of State. Blue Angel Inc., Charlottesville-1981.
84. Thompson, M., Ellis, R., & Wildavsky, A., (1990). *Cultural Theory* (Vol. 1, pp. 1–33). CO, Westview Press, Oxford.
85. Tolba, M. K., & El-Kholy, O. A., (1992). *The World Environment 1972–1992* (Vol. 1, pp. 1–44). Two decades of challenge, Chapman and Hall, London.
86. Tryzna, T. C., (1995). *A Sustainable World* (Vol. 1, pp. 1–54). Sacramento, IUCN.
87. Uddin, S. N., Taplin, R., & Yu, X., (2010). Towards a sustainable energy future—exploring current barriers and potential solutions in Thailand. *Environment, Development and Sustainability*, *12*(1), 63–87.

88. UNCED, (1992). *Report of the United Nations Conference on Environment and Development* (Vol. 1, pp. 1–33). (UNCED Report) A/CONF.151/5/Rev.1.

89. UNDP, (1999). *Human Development Report (1999)* (Vol. 1, pp. 1–33). New York. UNDP, (2000), Human Development Report (2000), New York.

90. Urbaniec, M., (2015). Towards sustainable development through eco-innovations: Drivers and barriers in Poland. *Economics and Sociology*, *8*(4), 179.

91. Von Schomberg, R., (2002). *The Objective of Sustainable Development, are we Coming Closer?* (Vol. 1, pp. 1–33). Foresight Working Papers Series No 1, https://papers.ssrn.com/sol3/papers.cfm?abstract_id=2436402.

92. Williamson, I., Ting, L., & Grant, D., (1999). The evolving role of land administration in support of sustainable development: A review of the United Nations-International Federation of surveyors Bathurst declaration for sustainable development. *Australian Surveyor*, *44*(2), 126–135.

93. Witkowski, K., (2017). Internet of things, big data, industry 4.0-innovative solutions in logistics and supply chains management. *Procedia Engineering*, *182*, 763–769.

94. Zelenika, I., & Pearce, J. M., (2014). Innovation through collaboration: Scaling up solutions for sustainable development. *Environment, Development and Sustainability*, *16*(6), 1299–1316.

INDEX

A

abandoned systems 115
agribusinesses 20
agricultural land 42, 45, 46
agroforestry 208
air pollution 48, 49, 58, 62, 63, 67
anthropological 164
Asian Development Bank 49
atmospheric pollution 180
average total cost curve (ATC 59

B

benchmarks 23
biodiversity 54, 113, 132
bioenergy 118, 121, 122, 132, 137
biological diversity 80
biological systems 84
biomass 71, 73, 91
biotechnology 15, 17, 18, 19, 20
business development 18
business environment 165
business networks 8
business ventures 76

C

carbon monoxide 83, 93
certification systems 122
chemical fertilizers 18
chemical hazards 55
civil society 52
climate change 15, 32
collective dynamics 147
Combating poverty 164
commercial energy 70, 71, 72, 78, 79, 80
communication technology 183
consumer demand 11
convenient energy 77
conventional fuels 76

critical thinking 3
crop development 15
Cultural Development 44

D

data arrangement 19
dematerialization 13, 32
Demography 164
Designated National Authority (DNA) 123
developed countries (DCs) 45, 70, 123, 198

E

eco-development strategy 150
ecological administration 5
Ecological Development 44
ecological economics 147, 148, 153
economic analysis 144
economic body 149
economic development 12, 13, 33
economic dynamics 144
economic growth 45, 47, 53, 115, 117
economic sustainability 116, 117
ecosystem 40
ecosystems 113, 114
electricity service 121
Employee satisfaction 167
Encouraging human settlement 164
energy administrations 87, 88, 89, 90
energy production systems 122
energy system 70, 74, 126, 127, 136
entrepreneurial support systems 118
environmental change 5, 8, 15, 22
Environmental degradation 47
Environmental destruction 47
environmental education 42
environmental framework 112
environmental issues 40, 42, 45, 46
environmental pollution 50, 65
environmental protection 160, 161
environmental quality 45, 53, 63
Equal distribution of wealth 197

F

financial development 88
financial institutions 118
financial instruments 14
financial systems 4
food security 15, 19, 20, 21
food systems 179, 186, 187, 188
fossil energy 92, 95
fossil fuel 81, 82, 83, 84, 85
fossil resources 87
fuel ignition 84

G

gas turbines 92, 93
gender equality 120
gender equity 15
genetically modified organisms (GMOs) 17
genetic modification 43
genuine progress indicator 7
geothermal 73, 91, 116, 117, 128, 132, 136
GHG emissions 116, 119, 122
global business 11
global climate 84
Global Environmental Outlook (GEO) 24
global income 53
global regulatory 9
global scale 46
Global Sustainable Development

176
global temperature 211
global warming 54
greenhouse gases 84, 91, 97
grid networks 118
Gross National Product (GNP) 197

H

healthcare 46
Health education 164
healthy nutrition 179
herbicides 18
horticultural 4, 15, 16, 17, 19, 20, 21
human beings 40, 44
human capital 208
human culture 26
Human Development 44, 62
human family 2
human health 56, 61
Human health protection 164
human innovativeness 26
human natural 10
human network 2, 24
human resources 13
human rights 114, 160
human skills 41
human-social-environmental systems 176
human system 26
hybrid arrangements 90
hydrocarbons 83, 84
hydropower 71, 73, 84, 91, 93, 114, 117, 127, 129, 132, 134
hydropower projects 114, 129
hydropower technology 114

I

industrial pollution 51, 53
industrial safety risks 11
information technologies (IT) 24
integrated gasifier combined cycle (IGCC) 93
Integrated Global Observing Strategy 22
intergenerational value 3
Intergovernmental Panel on Climate Change (IPCC) 22, 84, 211
International Council for Local Environmental Initiatives (ICLEI) 23, 31
international laws 55
international trade 46, 67, 68
International Union for Conservation of Nature (IUCN) 42, 196
investment systems 21
IPBES (Intergovernmental Science-Policy Platform on Biodiversity and Ecosystem Services) 183

K

Kuznets Model 51

L

labor market 169
landscape aesthetics 113
less developed countries (LDCs) 48
life support system 2, 5, 9, 11, 21
light water reactors (LWRs) 97
logistical obstructions 122

M

macroeconomic stability 120
marginal cost (MC) 59
market arrangement's 48
market-based energy 118
market penetration 122

military employments 97
millennium development goals (MDGs) 210
motor systems 89
multi-criteria analysis 113, 133
multi-criteria decision analysis 122
multidimensional 112

N

National sustainable development strategies (NSDS) 120
natural administration 5
natural capital 145, 148, 149
natural degradation 80
natural habitats 113
natural impression 2
natural resources 40, 45, 46, 52, 54, 56
Neoclassical analysts 145
nitrogen oxides 83
non-commercial energy 70, 71
Non-governmental organization 45

O

overseas development aid (ODA) 25

P

personal satisfaction 2, 10, 80
pesticides 18, 20
policy development 112
policymakers 183
Political Development 44
polygeneration 93
population growth 71, 76
Poverty Reduction Strategy 124
power generation 86, 88, 91, 93, 95, 96
prevailing system 123
psychological 116
public acceptance 114, 126, 141
purchasing power parities (PPP) 78

R

radioactive waste 86, 97, 98
radioactivity 10
refrigeration 70
renewable-based energy 92
renewable electricity technologies 125
renewable energy (RE) 45
renewable innovations 92
Renewable resources 87
resource degradation 49
rural development 165

S

Salary estimation alone 76
Satellite technologies 21
satisfactory energy 71
settled agriculture 43
small agricultural land 166
social challenges 192
social expenditure 116
social indicators 41
social infrastructure 8
social innovation 208
social policy 8
social prosperity 2
social structures 3
social systems 2
social welfare 40
socio-cultural issues 113
socio-economic 40

socio environmental 144
soil acidification 180
soil degradation 19
Soil erosion 54
Solar energy 41
solar irradiation 87
solar photovoltaics 92
solid waste management 14
Spiritual Development 44
strategic development 112
Stroller Chemical Co (SSC) 55
sulfur dioxide 51
Sustainability in International Law 2
sustainable advancement (SD) 2
Sustainable demographic dynamics 164
sustainable development goals (SDGs) 124, 179
sustainable energy systems 84

T

technical skills 115
technical support systems 115
term checking system 24
terrorism 194
thermonuclear fusion 98
transmitting ordinary 97
transportation 70, 72, 77, 80, 86, 88, 89, 95

U

under-developing countries (UDC) 48
underground water system 10
United Nations Conference on Environment and Development (UNCED) 43
United Nations Environment Program (UNEP) 42, 196
United Nations Office for Disaster Risk Reduction (UNISDR) 212

V

vehicular emission 51

W

watercourse diversion 43
water system 14, 16
wind generators 92
wind power 113, 134, 140
World Commission on Environment and Development (WCED) 42, 196
World Conservative Strategy 42
World Health Organization (WHO) 57
World Trade Organization 12, 38
worldwide organizations 86
worldwide surface temperature 23
World Wide Web 24